Structural Physiology of the *Cryptosporidium* Oocyst Wall

The mission of the Awwa Research Foundation (AwwaRF) is to advance the science of water to improve the quality of life. Funded primarily through annual subscription payments from over 1,000 utilities, consulting firms, and manufacturers in North America and abroad, AwwaRF sponsors research on all aspects of drinking water, including supply and resources, treatment, monitoring and analysis, distribution, management, and health effects.

From its headquarters in Denver, Colorado, the AwwaRF staff directs and supports the efforts of over 800 volunteers, who are the heart of the research program. These volunteers, serving on various boards and committees, use their expertise to select and monitor research studies to benefit the entire drinking water community.

Research findings are disseminated through a number of technology transfer activities, including research reports, conferences, videotape summaries, and periodicals.

Structural Physiology of the *Cryptosporidium* Oocyst Wall

Prepared by:
Honorine Ward, **Najma Bhat**, **Roberta O'Connor**, and **Smitha Jaison**
Tufts-New England Medical Center, Boston, MA

Giovanni Widmer, **Glenda Batzer**, **Betsy Corey**, **Barry Stein**, and **Saul Tzipori**
Tufts University School of Veterinary Medicine, North Grafton, MA

Joseph Crabb and **Shiva Gupta**
ImmuCell Corporation, Portland, ME

Jennifer L. Clancy
Clancy Environmental Consultants Inc., St. Albans, VT

and

Eric Osborne
Clayton County Water Authority, Morrow, GA

Sponsored by:
Awwa Research Foundation
6666 West Quincy Avenue, Denver, CO 80235-3098

Published by:

Awwa Research Foundation
Advancing the Science of Water®

IWA Publishing

DISCLAIMER

This study was funded by the Awwa Research Foundation (AwwaRF). AwwaRF assumes no responsibility for the content of the research study reported in this publication or for the opinions or statements of fact expressed in the report. The mention of trade names for commercial products does not represent or imply the approval or endorsement of AwwaRF. This report is presented solely for informational purposes.

Published by IWA Publishing, Alliance House, 12 Caxton Street, London SW1H 0QS, UK

Telephone: +44 (0) 20 7654 5500; Fax: +44 (0) 20 7654 5555; Email: publications@iwap.co.uk;
Web: **www.iwapublishing.com**

AwwaRF report number 91012F. Originally published by AwwaRF for its subscribers in 2004.
IWA Publishing version published 2005.

ISBN 1 84339 905 9

CONTENTS

TABLES

FIGURES

FOREWORD

The Awwa Research Foundation is a nonprofit corporation that is dedicated to the implementation of a research effort to help utilities respond to regulatory requirements and traditional high-priority concerns of the industry. The research agenda is developed through a process of consultation with subscribers and drinking water professionals. Under the umbrella of a Strategic Research Plan, the Research Advisory Council prioritizes the suggested projects based upon current and future needs, applicability, and past work; the recommendations are forwarded to the Board of Trustees for final selection. The foundation also sponsors research projects through an unsolicited proposal process; the Collaborative Research, Research Applications, and Tailored Collaboration programs; and various joint research efforts with organizations such as the U.S. Environmental Protection Agency, the U.S. Bureau of Reclamation, and the Association of California Water Agencies.

This publication is a result of one of these sponsored studies, and it is hoped that its findings will be applied in communities throughout the world. The following report serves not only as a means of communicating the results of the water industry's centralized research program but also as a tool to enlist the further support of the nonmember utilities and individuals.

Projects are managed closely from their inception to the final report by the foundation's staff and large cadre of volunteers who willingly contribute their time and expertise. The foundation serves a planning and management function and awards contracts to other institutions such as water utilities, universities, and engineering firms. The funding for this research effort comes primarily from the Subscription Program, through which water utilities subscribe to the research program and make an annual payment proportionate to the volume of water they deliver and consultants and manufacturers subscribe based on their annual billings. The program offers a cost-effective and fair method for funding research in the public interest.

A broad spectrum of water supply issues is addressed by the foundation's research agenda: resources, treatment and operations, distribution and storage, water quality and analysis, toxicology, economics, and management. The ultimate purpose of the coordinated effort is to assist water suppliers to provide the highest possible quality of water economically and reliably. The true benefits are realized when the results are implemented at the utility level. The foundation's trustees are pleased to offer this publication as a contribution toward that end.

Walter J. Bishop
Chair, Board of Trustees
Awwa Research Foundation

James F. Manwaring, P.E.
Executive Director
Awwa Research Foundation

ACKNOWLEDGMENTS

We wish to acknowledge Mr. Emerson Lomaquahu, Ms. Kathryn Martin, and Ms. Alice Fulmer, Awwa Resesarch Foundation project managers for administrative help and encouragement, as well as the Project Advisory Committee consisting of Drs. Anthony Lloyd, Huw Smith, Ramon Aboytes, and Jon Calomiris for helpful suggestions throughout this project. We thank Dr. Richard Nelson for DNA libraries, Dr. Marian Neutra for mAb CrA2, and Dr. Tony Smithyman for the Cellabs mAb. We wish to thank Dr. Herman van Halbeek and the University of California, San Diego, Glycotechnology Core facility for performing the high pH anion exchange chromatography analysis.

EXECUTIVE SUMMARY

The apicomplexan parasite *Cryptosporidium parvum* is a significant cause of human and animal diarrheal disease worldwide. This parasite is currently recognized as the causative agent of numerous outbreaks of waterborne diarrheal disease. *C. parvum* infection in immunocompetent individuals is asymptomatic or associated with self-limiting diarrheal illness. However in immunocompromised hosts, such as patients with acquired immunodeficiency syndrome (AIDS), *Cryptosporidium* may cause severe, protracted and possibly fatal diarrheal disease.

C. parvum isolates can be divided into two genetically distinct groups, one designated genotype I, exclusively associated with human infections, and the other genotype II, associated with both human and animal infections. The majority of infections associated with waterborne outbreaks are of genotype I. Published genotypic information of *C. parvum* from waterborne outbreaks particularly in the USA suggests that up to 80% of infected humans excrete genotype I oocysts. However, most studies related to water borne transmission use genotype II oocysts.

C. parvum oocysts can survive for many months in water and are resistant to several disinfectant treatments. The prolonged survival of oocysts as well as the resistance to disinfectants is attributed to the presence of a thick wall that is believed to serve a protective function by isolating the parasite from the external environment. Ultrastructurally, the oocyst wall consists of two electron dense layers, an outer irregular 10 nm layer separated by an electron-lucent space from an inner thicker electron dense layer. A distinctive feature of the oocyst wall is the presence of a suture spanning part of the circumference of the inner wall, which undergoes dissolution during excystation. Oocyst wall formation in *Cryptosporidium* is initiated in wall forming bodies present in macrogametes. Although the ultrastructural features of the oocyst wall and suture have been described in some detail, very little is known about the biochemical composition and structural physiology of these important structures. In addition, very little is known about the effect of various water treatment processes or disinfectants on individual components of the oocyst wall.

The integrity of the oocyst wall is responsible for prolonged survival of *C. parvum* in drinking water sources as well as its resistance to various disinfectants. The biochemical composition of specific components, which contribute to the structural integrity of the *Cryptosporidium* oocyst wall, and the effect of water treatment and purification processes on them are largely unknown. Knowledge of these components is therefore crucial in designing strategies directed at detecting and eliminating *C. parvum* from drinking water supplies.

RESEARCH OBJECTIVES

The overall goal of this project was to develop probes to identify and assess the relative roles of proteins, carbohydrates and lipids in the overall integrity and resistance of the *Cryptosporidium* oocyst wall for the assessment of disinfection of the oocyst.

The specific objectives were to

I. To develop probes to identify and assess the role of biochemical components in maintaining integrity of the oocyst wall.
   A. Protein components
      1. To identify proteins that are present in the oocyst wall and to determine the N-terminal amino acid sequence of selected proteins.
      2. To clone, sequence and express the genes encoding selected proteins and develop antibody probes for them.
   B. Carbohydrate components
      1. To identify and localize carbohydrate residues of oocyst wall components including the suture using lectins and glycosidases as probes.
      2. To determine the monosaccharide composition of oocyst wall components.
   C. Lipid components
      1. To characterize the presence of various classes of lipids in oocyst walls using fluorogenic probes.
      2. To determine the integrity of the oocyst wall and the presence of the lipid species after lipid extraction.

II. To identify and examine agents that would effectively dissolve, permeate, disrupt or breach the wall or suture and to determine their effect on biochemical components of the oocyst wall.
   A. Effect of disinfectant and other agents
      1. To study the effect of disinfectants and biochemical agents on oocyst wall integrity and on selected oocyst wall components
   B. Effect of water treatment processes
      1. To study the effects of water treatment processes on oocyst wall integrity and on selected oocyst wall components

APPROACH

The approach was to focus on a few well characterized molecules or reagents that have been shown to be particularly useful and specific for certain components, and use these molecules and reagents in conjunction with biochemical treatments known to affect certain classes of molecular structure. In this manner, the hope was to identify structures and/or molecules that have particular relevance to the unique resistance possessed by this organism. During the course of this project, the goal was to accomplish the objectives of developing probes such as antibodies for protein components, lectins for carbohydrate components and fluorogenic probes for lipid components. The use of these probes enabled the study of the effect of disinfectants such as chlorine and ozone on these components. However, the goals of cloning the gene encoding the protein recognized by the mAb 2C3/Cellabs and the study of the effects of water treatment processes on components of the oocyst wall could not be achieved. Some of these experiments are ongoing and will be pursued in future studies.

CONCLUSIONS

1. The antibodies cpRL3, mAb 2C3 and Cellabs mAb were used as probes for the oocyst wall proteins COWP and 2C3/Cellabs antigen. The results of these studies suggested that a) All three antibodies recognize oocyst wall proteins, with mAb 2C3 and Cellabs mAb seemingly recognizing the same epitopes, b) the cpRL3 epitope contains disulphide-linked subunits whereas the 2C3/Cellabs epitope does not, c) the 2C3/Cellabs epitope is not carbohydrate or carbohydrate-dependent in nature, and d) the cpRL3 epitope is present on the inner face of the oocyst, while the 2C3/Cellabs epitope is exposed on the outer face of the oocyst wall These antibodies are useful tools with which to probe the oocyst wall surface, and to correlate treatments that may change the integrity of the wall with differences in the presence of these molecules. These studies also indicate that the protein recognized by mAb 2C3/Cellabs is different from COWP. This is based on the findings that a) the molecular weights of the proteins as determined by immunoblotting are different, b) the 2C3 protein, unlike COWP does not contain inter or intra chain disulfide bonds, and c) the ultrastructural localization of the 2C3 protein is different from that of COWP. Attempts to clone the gene/s encoding the 2C3/Cellabs antigen/s were not successful.
2. A panel of lectins of varying specificity were used as probes to investigate the nature of carbohydrate residues present in the oocyst wall. Binding of FITC-conjugated lectins to intact oocysts of both genotype I and II, oocyst walls and SDS or hypochlorite treated oocyst walls was qualitatively assessed by microscopy. Reactivity of the lectins with fixed oocysts as well as intact oocysts in suspension was determined. The results indicated that lectins with specificity for GalNAc, galactose Gal, fucose, glucose, mannose and N-acetyl glucosamine reacted with oocysts and/or oocyst walls, suggesting the presence of these monosaccharides in oocyst walls. However, the pattern and intensity of reactivity varied with the conditions used and with the lectin. The strongest reactivity occurred with lectins specific for Gal and GalNAc. Binding of Gal/GalNAc-specific lectins HPA and MPA was confirmed and quantitated by flow cytometry. In contrast to the results obtained with fixed oocysts, intact oocysts reacted only with three lectins. The strongest reactivity by fluorescence occurred with MPA and AIA. A striking finding was that these lectins agglutinated oocysts and also induced excystation of intact oocysts in suspension.

   Glycosidases were also used as probes to confirm the lectin binding findings and to determine the type of glycosylation present. PNGase F and Endo-α-GalNAcase had no significant effect on lectin binding indicating that the GalNAc residues were not present in an N-glycosidic linkage and that they were not present as part of the Galβ1-3GalNAc determinant. Treatment with α-GalNAcase however resulted in significant reduction of binding of all the GalNAc-specific lectins, suggesting that these residues are terminal. The presence of large amounts of Glc, smaller amounts of galactosamine (presumably GalNAc before hydrolysis), galactose (Gal) mannose (Man), and glucosamine (GlcNAc) were qualitatively confirmed by HPAEC.
3. Fluorogenic lipid probes were used to analyze the classes of lipids present in oocyst walls. The results of these studies showed that intact oocysts and isolated oocyst walls failed to stain with the fluorogenic lipid probes tested. However, though non-treated, isolated oocyst walls did not incorporate NBD-C_6-HPC staining, SDS and heat-treated

oocyst walls brightly stained with NBD-C_6-HPC or NBD-C_6-sphingomyelin probes. Staining of oocyst walls that have been treated with SDS and heat indicate that phospholipids and sphingolipids are present in the oocyst walls whereas cholesterol is not. Moreover, it appears that the oocyst walls stain more brightly with NBD-C_6-sphingomyelin than with NBD-C_6-HPC, which suggests that the oocyst walls contain a higher ratio of sphingolipids to phospholipids. These data are consistent with a model of a tough outer layer present in *Cryptosporidium parvum* oocyst walls that must be denatured in order to incorporate any fluorescent lipid probes. For the three lipid classes examined, the following hierarchy appears to hold for the lipids of *C. parvum* oocyst walls: sphingolipids were present in greater amounts than phospholipids and cholesterol could not be detected.

4. *C. parvum* genotype I and genotype II oocysts were exposed to different doses of ozone and chlorine to investigate the effect of two disinfectants on the oocyst wall. Doses were chosen such that various levels of inactivation were achieved. Inactivation of genotype I and genotype II oocysts exposed to the highest ozone dose of 140 mg/min/l was confirmed in cell culture. The effect of ozone and chlorine (highest dose = 4000 mg/min/l) was probed using three methods: 1) reactivity with antibodies specific for oocyst wall antigens detected by mAbs 2C3/Cellabs and COWP, as detected by polyclonal antibody anti-cpRL3, 2) binding of two fluorescent lectin-FITC conjugates to oocyst carbohydrates, and 3) incorporation of a fluorescent phospholipid probe. Binding of the three types of fluorescent probes was assessed qualitatively by immunoblotting or by microscopy and quantitated by flow cytometry. Chlorine treatment reduced binding of the 2C3/Cellabs mAbs but not that of anti-cpRL3. Ozone had no detectable effect on 2C3/Cellabs or anti-cpRL3 binding. Binding of MPA to genotype I and genotype II oocysts was reduced by about 20% in response to ozone, whereas binding of HPA was decreased with genotype II oocysts but increased with type I. With respect to the uptake of labeled phospholipid, ozone and chlorine exposure resulted in the emergence of a brightly labeled oocyst subpopulation. The relative magnitude of this subpopulation was dose dependent and the fluorescence emitted by these oocysts approximated that emitted by heat inactivated oocysts. Together, these results suggest that 1) ozone and chlorine do not affect the integrity of the oocyst carbohydrates and 2) that these disinfectants permeabilize the oocyst wall, facilitating the uptake and incorporation of phospholipids into the oocysts.

RECOMMENDATIONS

The goal of this study was to determine the structural physiology of the *Cryptosporidium* oocyst wall as it relates to drinking water treatment. Based on this study, no immediate recommendations for drinking water treatment can be made. It is hoped that the information gained from this study can be used in the future to design specific strategies directed at detecting and eliminating *Cryptosporidium* from drinking water supplies.

FUTURE RESEARCH

Future studies will be directed at characterizing specific oocyst wall proteins, carbohydrates and lipids in greater detail, and determining their role in maintaining the structural physiology of the oocyst wall and the effect of disinfection and water treatment processes on these components.

CHAPTER 1
INTRODUCTION

BACKGROUND

The apicomplexan parasite *Cryptosporidium parvum* is a significant cause of human and animal diarrheal disease worldwide (O'Donoghue 1995; Guerrant 1997; Griffiths 1998; Tzipori and Griffiths 1998; Leav, Mackay and Ward 2003). *C. parvum* infection in immunocompetent individuals is asymptomatic or associated with self-limiting diarrheal illness. However in immunocompromised hosts, such as patients with acquired immunodeficiency syndrome (AIDS), *Cryptosporidium* may cause severe, protracted and possibly fatal diarrheal disease. This parasite has gained increasing notoriety as the causative agent of numerous outbreaks of waterborne diarrheal disease (Rose, Lisle and LeChevallier 1997; Fricker and Crabb 1998). In 1993, *C. parvum* was responsible for the largest ever documented outbreak of waterborne disease in the world, affecting an estimated 403,000 people in Milwaukee, WI, contributing to over 100 deaths (MacKenzie et al. 1994; Fricker and Crabb 1998) and resulting in estimated economic losses of over $96 million (Corso et al. 2003). Since then numerous outbreaks of waterborne cryptosporidiosis have been reported in the US and UK (Fricker and Crabb 1998).

Infection with *C. parvum* is initiated by ingestion of oocysts, which undergo excystation to release sporozoites (O'Donoghue 1995; Tzipori and Griffiths 1998). Sporozoites invade intestinal epithelial cells and undergo further intracellular development via asexual and sexual cycles resulting in the formation of oocysts which undergo sporulation within the host and are released into the external environment. *C. parvum* is highly infectious with as few as 30-100 oocysts required to initiate infection in healthy adult volunteers (DuPont et al. 1995). Several laboratories have shown that *C. parvum* isolates can be divided into two genetically distinct groups, one designated genotype I, exclusively associated with human infections, and the other genotype II, associated with both human and animal infections (Morgan et al. 1995; Bonnin et al. 1996; Carraway, Tzipori and Widmer 1996; Carraway, Tzipori and Widmer 1997; Morgan et al. 1997; Peng et al. 1997; Spano et al. 1997b; Spano et al. 1998). In addition to genetic differences, isolates of both genotype exhibit phenotypic differences such as animal infectivity (Widmer et al. 1998). Of relevance to the water industry is the observation that the majority of infections associated with waterborne outbreaks are of genotype I (Tzipori and Widmer 2000) (Fayer, Morgan and Upton 2000). Published genotypic information of *C. parvum* from waterborne outbreaks suggest that up to 80% of infected humans excrete genotype I oocysts (Tzipori and Widmer 2000).

C. parvum oocysts can survive for many months in water and are resistant to several disinfectant treatments (Campbell et al. 1982; Korich et al. 1990). The prolonged survival of oocysts as well as the resistance to disinfectants is attributed to the presence of a thick wall that is believed to serve a protective function by isolating the parasite from the external environment. Ultrastructurally, the oocyst wall consists of two electron dense layers, an outer irregular 10 nm layer separated by an electron-lucent space from an inner thicker electron dense layer (Reese et al. 1982; Current and Reese 1986; Ungar 1990). A distinctive feature of the oocyst wall is the presence of a suture spanning part of the circumference of the inner wall, which undergoes dissolution during excystation. Oocyst wall formation in *Cryptosporidium* is initiated in wall forming bodies present in macrogametes (Current and Reese 1986). Although the ultrastructural features of the oocyst wall and suture have been described in some detail, very little is known about the biochemical composition and structural physiology of these important structures. In addition, very little is known about the effect of various water treatment processes or disinfectants on individual components of the oocyst wall. The current state of knowledge about various biochemical components of the oocyst wall are summarized below.

Proteins

Cryptosporidium oocyst walls have been reported by different investigators to contain a number of proteins with a wide range of molecular weights (MW). Lumb, Lanser and O'Donoghue (1988) identified 5 major proteins 15-96 kDa in molecular weight by ^{125}I surface labeling of intact oocysts. Surface labeling of oocysts by Tilley et al. revealed a number of bands 32-290 kDa in molecular weight (Tilley et al. 1990). Lumb et al. also reported the presence of 7 major bands of 14-190 kDa in MW by SDS PAGE in isolated oocyst walls (Lumb, Lanser and O'Donoghue 1988). SDS PAGE of isolated oocyst walls by Anthony et al. revealed 7 major bands 14-145 kDa in MW (Anthony et al. 1997). Tilley identified 8 major bands in Percoll purified oocyst walls (Tilley and Upton 1997). The reasons for discrepancies in the number and MW of oocyst wall proteins previously reported are unknown but may be attributable to: a) differences in the isolation and purification procedures for obtaining oocysts and oocyst walls, b) differences in labeling or staining procedures, c) presence of contaminating sporozoite proteins, or d) isolate differences.

The best-characterized oocyst wall protein is a 190 kDa protein named COWP (*Cryptosporidium* oocyst wall protein) which was initially described by two different groups (Lally et al. 1992; Ranucci et al. 1993). Partial genomic DNA clones encoding this protein were identified and sequenced, initially by Lally et al. (1992) and subsequently by Ranucci et al. (1993). Subsequently the latter group cloned and sequenced a full-length genomic clone (Spano et al. 1997a). The COWP gene consists of an open reading frame of 4866 nucleotides coding for a protein of 1622 amino acids with a predicted molecular weight of 174 kDa for the mature protein. Analysis of the amino acid sequence revealed the presence of a putative signal peptide, 9 potential N-linked glycosylation sites and unusually high contents of cysteine, proline and histidine. There are two different repeat motifs; both characterized by the presence of cysteine residues at conserved positions. The function of these repeat units is unknown, although it has been speculated that because of their cysteine content they may have a role in stabilizing protein structure through disulphide bonds. COWP is present as a single copy gene and appears to be conserved among various human and bovine *C. parvum* isolates. Polyclonal and monoclonal antibodies raised to a recombinant protein encoded by a portion of the gene recognized a band of 190 kDa in oocyst lysates by immunoblotting and reacted with the oocyst wall by confocal immunofluorescence microscopy (Ranucci et al. 1993). By immunoelectron microscopy, COWP was localized to the inner layer of the oocyst wall as well as to large vesicular cytoplasmic inclusions and smaller Type I wall-forming bodies of macrogametes (Spano et al. 1997a).

Carbohydrates

The presence of carbohydrates in oocyst wall components has been suggested by studies of lectin-binding to intact oocysts (Llovo et al. 1993; Gut and Nelson 1994). Llovo et al. showed that GlcNAc-specific lectins agglutinated intact oocysts, suggesting that this sugar moiety is present on the oocyst surface (Llovo et al. 1993). Gut and Nelson reported reactivity of FITC-labeled GalNAc-specific lectins with the surface of oocysts (Gut and Nelson 1994). Anthony et al., reported the presence of Gal, Glc, Man and talose in oocyst walls by GC-MS and HPLC analysis of purified oocyst walls (Anthony et al. 1997). A recent study described the presence of a "glycocalyx" surrounding the oocyst as determined by electron microscopy of ruthenium red stained oocysts (Nanduri et al. 1999). Carbohydrate compositional analysis of this material revealed the presence of GalNAc, Glc, Gal, Man, xylose, and ribose.

Lipids

The lipid composition of intact oocysts was studied by Mitschler, Welti and Upton who showed that phosphatidyl choline was the predominant phospholipid, with smaller amounts of sphingomyelin (Mitschler, Welti and Upton 1994). Cholesterol was the only sterol detected. Major fatty acids included palmitic (16:0), stearic (18:0), oleic (18:1) and linoleic (18:2) acids. Subsequently, purified oocyst walls were analyzed by GC-MS for fatty acid composition, and by TLC for glycolipid composition (Anthony et al. 1997). Major fatty acids that were identified as fatty acid methyl esters after chloroform-methanol extraction and methanolysis included C6, C8, 9 octadecenoic and 11,14 eicosadienoic acids. In addition, several aliphatic hydrocarbons were identified in hexane and chloroform-methanol extracts, some of which have high melting points, suggestive of the waxy constituents found in mycobacterial cell walls (mycolic acids) that impart the acid fast staining character to these cells. These authors reported the absence of detectable phospholipids.

OBJECTIVES OF THE STUDY

The integrity of the oocyst wall is responsible for prolonged survival of *C. parvum* in drinking water sources as well as its resistance to various disinfectants. The biochemical composition of specific components, which contribute to the structural integrity of the *Cryptosporidium* oocyst wall, and the effect of water treatment and purification processes on them are largely unknown. Knowledge of these components is therefore crucial in designing strategies directed at detecting and eliminating *C. parvum* from drinking water supplies.

The overall goal of this project was to develop probes to identify and assess the relative roles of proteins, carbohydrates and lipids in the overall integrity and resistance of the *Cryptosporidium* oocyst wall for the assessment of disinfection of the oocyst. It was not feasible to systematically dissect the oocyst wall with the goal of identifying the relative position and abundance of each and every molecular component in a short time and with limited resources, Therefore, our approach was to focus on a few well characterized molecules, or reagents that have been shown to be particularly useful and specific for certain components, and use these molecules and reagents in conjunction with biochemical treatments known to affect certain classes of molecular structure. In this manner, we hoped to identify structures and/or molecules that have particular relevance to the unique resistance possessed by this organism. These structures and/or molecules can be characterized in detail in future studies. Knowledge gained from these studies will be of value to water utilities and regulatory agencies in order to improve current approaches to water purification and to design new strategies.

The specific objectives of this project were:

I. To develop probes to identify and assess the role of biochemical components in maintaining integrity of the oocyst wall.
   A. Protein components
      1. To identify proteins that are present in the oocyst wall and to determine the N-terminal amino acid sequence of selected proteins.
      2. To clone, sequence and express the genes encoding selected proteins and develop antibody probes for them.
   B. Carbohydrate components
      1. To identify and localize carbohydrate residues of oocyst wall components including the suture using lectins and glycosidases as probes.
      2. To determine the monosaccharide composition of oocyst wall components.
   C. Lipid components
      1. To characterize the presence of various classes of lipids in oocyst walls using fluorogenic probes.
      2. To determine the integrity of the oocyst wall and the presence of the lipid species after lipid extraction.

II. To identify and examine agents that would effectively dissolve, permeate, disrupt or breach the wall or suture and to determine their effect on biochemical components of the oocyst wall.
   A. Effect of disinfectants and other agents
      1. To study the effect of disinfectants and biochemical agents on oocyst wall integrity and on selected oocyst wall components
   B. Effect of water treatment processes
      1. To study the effects of water treatment processes on oocyst wall integrity and on selected oocyst wall components

CHAPTER 2
MATERIALS AND METHODS

OOCYSTS

Oocyst Isolates

In this project the following oocysts isolates were used as representatives of genotype II: GCH1, originally obtained from a human AIDS patient and passage in neonatal calves (Tzipori et al. 1994), Iowa (Mead et al. 1990; Ortega et al. 1991), UCP, originally obtained from Dr. Beth Ungar at USUHS isolated from a calf, and passaged in neonatal calves since 1989, and MD isolates, originally isolated from a red deer calf *(Cervus elaphus)* in 1986; the isolate was obtained from S. Wright, Moredun Research Institute, Edinburgh, Scotland and has been passaged in bovine calves since 1999 at the University of Arizona (Okhuysen et al. 2002). NEMC1 (Widmer et al. 2000) and TU502, previously designated UG502 (Akiyoshi et al. 2002), were used as a representative genotype I isolate.

Oocyst Propagation

GCH1, UCP and Iowa isolates were propagated in neonatal calves as previously described (Tzipori et al. 1994). TU502 was propagated in gnotobiotic piglets (Akiyoshi et al. 2002), and *C. parvum* isolate MD was propagated in interferon-γ knock-out mice (Dalton et al. 1993; Theodos et al. 1997).

Oocyst Purification

Oocysts were purified from fecal suspensions within 2 weeks of propagation. Since procedures such as ether extraction and hypochlorite treatment may result in biochemical alterations to oocyst wall components, use of these procedures in the purification process were eliminated or minimized. For most experiments, oocysts were purified by salt flotation followed by Percoll density gradient centrifugation (Arrowood and Sterling 1987). For the disinfection experiments, oocysts were purified first by salt floatation followed by sedimentation on a 15%/25% Nycodenz (Sigma Chemical Co.) step gradient (Widmer et al. 1998). The final purification was performed by IMS (CryptoScan, ImmuCell, Portland, ME). Oocysts less then 2 months old were used.

Oocyst Wall Preparation

Intact oocysts (unbleached) were resuspended in PBS and excysted by incubation at 37°C for 2 hours. Oocyst walls were isolated from the mixture of excysted and intact oocysts by Percoll density gradient centrifugation (Arrowood and Sterling 1987). The preparation contained >95% oocyst walls as determined by phase contrast microscopy. Aliquots of isolated oocyst walls were boiled with 1% SDS for 10 min. Proteins eluted from the walls by this treatment were saved.

ANTIBODIES

mAb 2C3

mAb 2C3 was generated at ICC as follows. BALB/cJ mice were immunized with oocyst walls in Freund's complete and incomplete adjuvant. Boosts were accomplished with oocyst walls or whole oocysts (formalin fixed). Hybridomas were generated using standard methods and clones screened by IFA using formalinized whole oocysts. After 10-20 fusions, a strongly fluorescing mAb, designated 2C3, which recognized the surface of the oocyst in a uniform fashion, was obtained.

Cellabs mAb

The Cellabs murine IgG mAb to oocyst wall protein/s was obtained as a kind gift from Dr. A.M. Smithyman, Cellabs Pty Ltd.

Anti-cpRL3 Polyclonal Antibody

Clone cpRL3 was provided to ICC by Dr. A. Crisanti, Imperial College of Science, Technology and Medicine, London, UK. This clone encodes a polypeptide constituting the amino terminal 786 aa of COWP (Ranucci et al. 1993), and the insert was ligated into pUA1/pDS56/RBSII-E^{-}, a 6 × his construct. This permitted growth of the M15 strain carrying this plasmid, induction with IPTG, and purification of this recombinant using nickel-affinity chromatography. Immunization of rabbits with 0.1 to 0.5 mg of this recombinant with Freund's complete and incomplete adjuvant resulted in sera that recognized the ~86 kDa polypeptide used to immunize, and also recognized oocysts in immunofluorescence (IF) experiments (described below).

IMMUNOFLUORESCENCE MICROSCOPY

Immunofluorescence assays were performed using mAb 2C3, Cellabs mAb and anti-cpRL3. mAb 2C3 was FITC labeled. Anti-cpRL3 IgG was prepared using protein-A affinity chromatography by standard methods (Goding 1986). Intact oocysts, as well as isolated walls (10^5/well), were placed on poly-L-lysine-coated multi-well slides (Falcon) and air-dried. Slides were fixed with 4% paraformaldehyde in PBS to visualize surface components. Slides were washed with PBS and stored at –20°C until use. After equilibration with PBS at room temperature, non-specific binding was blocked with 1% normal goat serum in PBS for 1 hr at room temperature. Slides were then incubated with antibodies diluted in 1% normal goat serum in PBS for 1 hour at room temperature. They were then washed three times with PBS, incubated with FITC-conjugated goat anti-mouse or anti-rabbit IgG (Sigma Chemical Co.) for 1 hour at room temperature. After three washes with PBS slides were mounted with Vectashield (Vector Laboratories Inc, Burlingame, CA) and viewed by epifluorescence and phase contrast microscopy. Double-label experiments were conducted by co-incubating oocysts, in suspension, with FITC 2C3 and rabbit anti-cpRL3 IgG, followed by centrifugation, washing in PBS, and re-incubating the mixture with TRITC-goat anti-rabbit IgG (Sigma Chemical Co.). Resulting labeled suspensions were visualized using Hoffman Modulation Contrast, FITC and TRITC optics.

ONE- AND TWO-DIMENSIONAL SDS PAGE

Aliquots (10^6 oocyst equivalents per lane) of intact oocysts, oocyst walls, SDS-eluted proteins and SDS-treated oocyst walls were subjected to 5-15% gradient SDS PAGE using the Laemmli buffer system as previously described (Harlow and Lane 1988). To determine the effect of reducing agents on the oocyst wall antigens, isolated oocyst walls were boiled with sample buffer containing 20% 2-mercaptoethanol (2-ME) for 5 min or treated with 40 mM dithiotreitol (DTT) for 18 hours at 56°C.

For 2-D gel electrophoresis, ocysts or oocyst walls were solubilized in sample buffer containing 9 M urea (W/V), 2% (V/V) NP-40, 2% (V/V) ampholites 3-10 (Sigma Chemical Co.) and 5% (V/V) 2- mercaptoethanol. Insoluble material was removed by centrifugation at 14,000 × g for 10 min. First dimensional electrophoresis was performed in 7.5 cm × 2 mm glass tubes. A gel solution containing 9 M urea (W/V), 2% (V/V) ampholites, 3.3%(V/V) bis-acrylamide, 2% (V/V) NP-40, 10% (V/V) TEMED and 10% (W/V) ammonium persulfate, was poured into the tubes and allowed to polymerize at RT for 45 min. A 5 µl sample overlay solution containing 9 M urea and 2% ampholites was pre-run for 30 min at 200 volts with cooling and then a 10 µl protein sample was applied to the gel. Rabbit muscle CPK solution (5 mg/ml) in 8 M urea solution containing 1% 2-mercaptoethanol was used as reference for isoelectric focusing. The gel was run for 3 hours at 400 volts with cooling followed by 500 volts for one hour and then equilibrated in 0.125 Tris -HCl pH 6.8 containing 10% glycerol, 2% SDS, 4.9 mM dithiothreitol and 1% bromophenol blue (equilibration buffer) for 1 hour. Second dimensional SDS-PAGE was performed using a 5-15% gradient as described above. The tube gel was carefully transferred on to the top of the slab gel and sealed with 0.5% molten agarose in 0.125 M Tris-HCl buffer pH 6.8 containing 0.02% SDS. The gel was run at constant current (20 mA/gel) with cooling and resolved proteins visualized by silver staining.

IMMUNOBLOTTING

Immunoblotting of Oocysts and Oocyst Walls

Oocyst or oocyst wall proteins were resolved by SDS PAGE and transferred to nitrocellulose as described previously (Harlow and Lane 1988) at 300 mAmps constant current for 2 hours at room temperature or overnight at 4°C and probed with a 1:100 dilution of mAb 2C3 culture supernatant, Cellabs mAb 1-10 µg/ml, anti-cpRL3 culture supernatant 1:100 or IgG, 1-10 µg/ml. Anti-mouse-IgG or anti-rabbit-IgG conjugated to HRP (Pierce Chemical Co., Rockford, IL) were used as secondary antibodies. Bound proteins were detected by chemiluminescence using the SuperSignal system (Pierce Chemical Co.).

Periodate Acid Treatment of Immunoblots

Isolated oocyst walls were subjected to 5-15% gradient SDS PAGE, transferred to nitrocellulose and strips rinsed with 20 mM sodium acetate, pH 4.5. They were then incubated in the same buffer for 1 hr at room temperature, followed by incubation in 5 mM periodic acid in 20 mM sodium acetate, pH 4.5 for 1 hr at room temperature in the dark. After rinsing with 20 mM sodium acetate, pH 4.5, strips were incubated with 50 mM sodium borohydride in PBS for 30 min at room temperature. Control strips were not incubated with periodate acid. Strips were then probed with mAb 2C3 and as a positive control, mAb 4E9, which recognizes a carbohydrate epitope present on the *C. parvum* glycoproteins GP900 and gp40.

DNA LIBRARY SCREENING

Approximately 5×10^6 to 10^7 phages from each of a λzap II Iowa sporozoite cDNA library, a λgt 11 oocyst (NINC isolate) genomic library, both from Dr. Richard Nelson, UCSF, as well as a λZAP Express GCH1 oocyst genomic library (partial Sau3AI digest) and a λZAP II GCH1 oocyst genomic library (complete *Eco*RI digest), both constructed at TUSVM, were screened as described previously (Sambrook and Russell 2001) with supernatant from a 2C3 producing hybridoma as well as with CrA2 (an IgA mAb to gp15, a *C. parvum* surface glycoprotein, produced at TUSVM in collaboration with Dr. Marian Neutra, Children's Hospital Boston, MA) and 8A9.2, (a mouse α-sporozoite IgG mAb antibody developed at TUSVM). Sonicated oocysts were used as a positive control antigen preparation to ensure that the antibodies and detection systems worked as expected. Both antibodies detected the controls. The SuperSignal chemiluminescent detection system (Pierce Chemical Co.) was used in these experiments.

FLOW CYTOMETRY-BASED ANTIBODY BINDING

Oocysts were labeled with mAb 2C3 followed by FITC-conjugated goat-anti-mouse IgG (Sigma Chemical Co.) and binding analyzed by flow cytometry. Oocyst fluorescence and scatter signals were analyzed on a Coulter Epics flow cytometer. FITC fluorescence was acquired on the FL1 channel. Since most analyses were performed on purified oocysts, no gates were applied. Signals from small particles were eliminated using a discriminator on the forward scatter signal. Typically, 50,000 events were acquired. Representative settings for data acquisition were as follows:

Signal	Volts	Gain
Forward scatter (FS)	50	10
Side scatter (SS)	100	50
Fluorescence 1 (FL1)	491	10

FL1 settings were adjusted according to fluorescence intensity.

FLUORESCENCE MICROSCOPY-BASED LECTIN BINDING

Lectin Binding to Fixed Oocysts

Lectin binding experiments were performed on fixed intact oocysts, with and without prior hypochlorite treatment (1.75% sodium hypochlorite for 10 min on ice) as well as on isolated oocyst walls with and without prior SDS treatment (1% SDS for 10 min at 100°C). Intact oocysts as well as isolated walls (10^5/well) were placed on poly-L-lysine-coated multi-well slides and air-dried. Slides were fixed with 4% paraformaldehyde in PBS to visualize surface components. Slides were washed with PBS and stored at –20°C until use. After washing with PBS, slides were incubated with a panel of FITC-conjugated lectins (EY Laboratories Inc, San Mateo, CA) (Table 3.1) at 1 and 10 μg/ml in PBS for 1 hour at room temperature. They were then washed three times with PBS, mounted with Vectashield (Vector Laboratories, Inc.) and viewed by epifluorescence and phase contrast microscopy. Specificity of binding was determined by pre-incubation of the lectins with their cognate monosaccharide for 30 min prior to the assay.

Lectin Binding to Unfixed Oocysts in Suspension

Oocysts (5×10^6 per reaction) were incubated with 100 μl each of 12 FITC-conjugated lectins of varying specificity (10 μg/ml) in PBS for 1 hour at 4°C. They were then washed three times with PBS by microcentrifugation at 4°C and resuspended in a final volume of 25 μl. 10 μl was placed on a glass slide, a cover slip applied and the preparation viewed by epifluorescence and phase contrast microscopy as soon as possible to avoid fading of the fluorescence and excystation of oocysts. These studies were performed on 2 different occasions using the same batch of oocysts of each genotype.

FLOW CYTOMETRY-BASED LECTIN BINDING

Oocysts were labeled with FITC-conjugated lectins either directly or following fixation in 4% paraformaldehyde in PBS as described in the previous "lectin binding to fixed oocysts" section. Lectins were reacted with the oocysts overnight at 4°C and unbound lectin removed by washing twice in PBS. Flow cytometry was performed as described in the previous "flow cytometry-based antibody binding" section.

GLYCOSIDASE DIGESTION

Whole oocysts and oocyst walls (10 μl of a suspension of 10^7/ml per well) were air dried and fixed with 4% paraformaldehyde in PBS or methanol for 10 min at room temperature, followed by washing in PBS for 5 min. Slides were treated with the following glycosidases for 18 hours at 37°C. As controls, slides were treated with the buffer alone at the same time under the same conditions.

1. PNGaseF (New England BioLabs, Beverly, MA) 10 mU/ml in buffer containing 20 mM sodium phosphate, pH 7.5, 50 mM EDTA
2. Endo-α-GalNAcase (New England BioLabs) 40 mU/ml in buffer containing 100 mM sodium phosphate/sodium citrate, pH 6.0
3. α-GalNAcase (New England BioLabs) 5 mU/ml in buffer containing 100 mM sodium phosphate/sodium citrate, pH 4.0

HIGH pH ANION EXCHANGE CHROMATOGRAPHY

Isolated oocyst walls were analyzed for monosaccharide composition following acid hydrolysis by high pH anion exchange chromatography with pulsed amperometric detection (HPAEC-PAD) as described previously (Manzi and Halbeek 1999) at the UCSD Glycotechnology Core facility.

FLUORESCENCE MICROSCOPY-BASED LIPID PROBE BINDING

The lipid probes (Molecular Probes, Eugene, OR) described in Table 3.9 were used. All probes except the oleic acid analogue were dissolved in ethanol at stock concentrations of 1 mg/ml. The oleic acid analogue was dissolved at the same concentration in acetonitrile.

Staining With Fluorogenic Probes

For staining, 100 μl of oocyst suspension was diluted to 1 ml in PBS in an Eppendorf microcentrifuge tube. 20 μl of desired probe was added and vortexed until sufficiently dispersed through solution. Initially 12-(9-anthroyloxy) oleic acid precipitated out of solution and attached to the sides of the plastic microcentrifuge tubes. In order to get this probe into solution ~10 μl of 1% Triton-x100 was added with mild vortexing. The probe was then added while gently vortexing. Various incubation times for staining, ranging from 15 to 90 minutes were tested, but the exposure time did not seem to have an effect on the quality or intensity of the fluorescence. Therefore 30 minutes was used for all of the following staining experiments. After incubation, the solutions were centrifuged at 10,000 × g for 5 minutes. The supernatant was aspirated off and the pellet resuspended in 100 μl of PBS.

In order to confirm that the probes were functional, cultures of Madin-Darby Bovine Kidney cells (MDBK) were prepared in Lab-Tek® Chamber Slides (Falcon, Franklin Lakes, NJ). The cells were maintained in DME-PSG medium. The probe was added in dilutions of Dulbeco's PBS. All five of the probes fluoresced brightly with MDBK cells, indicating the presence of the standard lipid classes known to be present in eukaryotic cells.

Treatments and Conditions

For freeze-thaw treatment, oocysts were frozen at –80°C, followed by thawing. For probing at elevated temperatures, three Eppendorf tubes of UCP oocysts (1.0 ml/tube; 4.0 × 10^6 oocysts/ml) were spun at 10,000 × g for 10 minutes. The isolated pellets were resuspended in 1× PBS (50 μl, and 1 μl of NBD-C6-HPC (1 mg/ml) was added to each solution. The solution was incubated at room temperature, 37°C, and 50°C for 30 minutes. After incubating, the preparation was diluted in PBS (1.0 ml total volume). Again, pellets were isolated by spinning each of the solutions at 10,000 × g for 10 minutes, and resuspended in PBS (10 μl). Slides were prepared with 5 μl of each solution, and then covered and sealed with cover slips and nail polish. For hypochlorite treatment, aliquots of oocysts were treated to bring the final hypochlorite concentration to 1% v/v. The solution was vortexed for ~1 minute and incubated at room temperature for 30 minutes. Time trials showed little difference in fluorescent patterns between oocysts exposed to hypochlorite for 15, 30, and 90 minutes. After incubation, solutions were centrifuged at 10,000 × g for 5 minutes, washed twice in Milli-Q water (MQW), and stained.

Oocyst Walls

Combined SDS and heat treatments were performed on oocyst walls and intact oocysts. GCH1 oocyst wall preparation (300 μl; 2.0 × 10^7 walls/ml) was combined with MQW (850 μl) in an Eppendorf tube, and spun at 10,000 × g for 10 minutes. The isolated pellet was resuspended in 1% SDS (50 μl), and incubated for 10 minutes in a boiling water bath. After the SDS and heat treatment, the solution's volume was brought up to 1.0 ml in MQW, and then spun down at 10,000 × g for 10 minutes. The pellet was then resuspended in PBS (150 μl), and the solution was divided into three, 50-μl aliquots. Three different probes were tested (1 μl): NBD-C_6-HPC (1 mg/ml), NBD-C_6-sphingomyelin (1 mg/ml), or NBD-Sterol (1 mg/ml). Each probe was added to one aliquot of oocyst walls suspended in PBS, and incubated at room temperature in the dark for 30 minutes. The preparation was then diluted in PBS (1.0 ml total volume). Again, pellets were isolated by spinning each of the solutions at 10,000 × g for 10 minutes, and resuspended in PBS (10 μl). Slides were prepared with 5 μl of each solution, and then covered and sealed with cover slips and nail polish. Process controls for each type of fluorescence were prepared in the same manner except that they were not subject to SDS and heat treatment.

Whole Oocysts

Two Eppendorf tubes of UCP oocysts (1.0 ml/tube; 4.0 × 10^6 oocysts/ml) were spun at 10,000 × g for 10 minutes. The isolated pellet was resuspended in 1% SDS (50 μl), and incubated for 10 minutes in a boiling water bath. After the SDS and heat treatment, the solution's volume was brought up to 1.0 ml in MQW, and then spun down at 10,000 × g for 10 minutes. The pellet was then resuspended in PBS (50 μl). Two probes were used: NBD-C_6-HPC (1 mg/ml) or NBD-Sterol (1 mg/ml). Each probe was added to one aliquot of oocysts suspended in PBS, and incubated at room temperature without light for 30 minutes. The fluorescence was then diluted in PBS (1.0 ml total volume). Again, pellets were isolated by spinning each of the solutions at 10,000 × g for 10 minutes, and resuspended in PBS (10 μl). Slides were prepared with 5 μl of each solution, and then covered and sealed with cover slips and nail polish. Controls for each type of fluorescence were prepared in the same manner except that they were not subject to SDS and heat treatment.

Solvent Extractions

Extractions were conducted with ethanol, hexane, chloroform:methanol (2:1), 0.4 M sodium acetate methanol:chloroform:water (60:30:8) and acetone. Control samples were subjected to SDS treatment and all procedural steps, except the extraction solvent wash. Extractions with hexane, ethanol and chloroform:methanol were initially conducted with UCP oocysts and then repeated with GCH1 oocysts. For extractions following SDS treatment, aliquots of 200-500 μl of oocysts were treated with 1% SDS and boiled for 10 minutes. After one wash in PBS, the pellet was resuspended and diluted to 1 ml in extraction solvent. Oocyst extraction solutions were vortexed for ~7 minutes, then centrifuged for 5 minutes at 10,000 × g. The pellet was washed in 1 ml of 1× PBS, resuspended and stained with probe.

Microscopy of Stained Preparations

Initial screening was conducted on a Nikon DIAPHOT-TMD inverted microscope with epi-fluorescence, a Nikon PlanDL 40× objective with a 0.70 numerical aperture, and a Nikon Eplan 100× oil immersion objective with a 1.25 numerical aperture. An Olympus BX-60 compound microscope with Normarski Differential Interference Contrast optics, epi-fluorescence, and an Olympus UplanFluor 100× oil immersion objective with a 1.30 numerical aperture were used to view the isolated GCH1 oocyst walls and UCP whole oocysts that were treated with SDS and heat. Photomicrography of the GCH1 oocyst walls and UCP whole oocysts was carried out using a Diagnostic Instruments Spot II digital camera and Diagnostic Instruments Spot v. 3.1 for Windows software, which was run on a Dell Optiplex 450 GX1p Pentium II (courtesy of Dr. James Weber, University of Maine at Orono).

FLOW CYTOMETRY-BASED LIPID PROBE BINDING

For lipid labeling experiments, oocysts were boiled in 1% SDS for 10 min. Unboiled controls for all samples were incubated in parallel at room temperature. In some experiments oocyst walls were used. A stock solution of 5 mg/ml NBD-C_6-HPC (Molecular Probes) in ethanol was diluted to 1 mg/ml in PBS. Following boiling, oocysts were washed once in PBS and resuspended in 50 µl PBS containing 40 µg/ml 2-(6-(7-nitrobenz-2-oxa-1,3-diazol-4-yl)amino) hexanoyl-1-hexadecanoyl-sn-glycero-3-phosphocholine (NBD-C_6-HPC). Oocysts were incubated in staining solution for 30 min at room temperature in the dark. Dye uptake by oocysts was confirmed microscopically. Thereafter oocysts were washed once with PBS and resuspended in 500 µl PBS for flow cytometry, as described in the previous "flow cytometry-based antibody binding" section.

ELECTRON MICROSCOPY

Immunoelectron Microscopy

Samples of *C. parvum* oocysts (GCH1) or the terminal ileum of *C. parvum* infected IFNγ knockout mice (Griffiths et al. 1998) were fixed in 0.25% glutaraldehyde, 4% paraformaldehyde in 0.1 M sodium phosphate buffer, pH 7.2 and processed for immunoelectron microscopy as described previously (Cevallos et al. 2000), using mAb 2C3 and colloidal gold-labeled anti-mouse IgG as primary and secondary antibodies respectively.

Scanning Electron Microscopy

Treated and untreated oocysts were processed in parallel. Oocysts were processed according to standard procedures as described (Bozzola 1999). Oocyst suspensions were fixed in 2% glutaraldehyde in 0.1 M sodium cacodylate buffer, pH 7.2 for 1 hr at room temperature. Poly-L-lysine coated coverslips were prepared by placing coverslips on drops of 1% poly-L-lysine for 5 min and washing with distilled water. Oocyst suspensions were placed on the coated coverslips, allowed to stand for 5 min and washed with distilled water. Samples were dehydrated in a graded ethanol series, critical point dried with CO_2, sputter coated with gold, viewed, and photographed using a JEOL 840 or JEOL 6020FE microscope.

OZONE TREATMENTS

Ozone treatments were performed at Clancy Environmental Consultants. Exposures were performed in phosphate buffer at pH 7 and 20°C. Initial ozone concentration was set at 1.4 mg/l, and ozone residual monitored at different time points during the experiment. The ozone dose was expected, on the basis of previous observations (AwwaRF project 492), to result in complete or nearly complete oocyst inactivation. Following exposure to ozone, the reactions were quenched with sodium thiosulphate and the oocysts were returned to TUSVM by overnight delivery. Based on previous observations (AwwaRF project 492) (Widmer et al. 2002) process controls were run in duplicate, quenched and unquenched, to control for the effect of sodium thiosulphate on the oocysts.

CHLORINE TREATMENTS

Chlorine exposures were performed at Clancy Environmental Consultants. Laboratory grade deionized water was adjusted to pH 5 with 0.01 N HCl and dosed to approximately 4 mg/l free available chlorine with sodium hypochlorite. One eighth of the oocyst stock was set aside as trip control. Oocysts were suspended in 10 mM phosphate buffer at pH 5. For the two process controls, ⅛ each of the oocyst stock was resuspended in phosphate buffer and stirred at room temperature for the duration of the exposure (16.7 hr). The treatments were 250 min exposure to 4 mg/l chlorine, 500 min exposure and 1000 min exposure. The 500 min and 1000 min exposure samples were each seeded with ¼ of the oocyst stock to compensate for the expected loss of oocysts. Chlorine demand of the oocyst suspension was determined prior to conducting the exposure trials. A portion of 100 ml of chlorine solution was transferred to a glass beaker with a teflon stir bar. A portion of 244 µl of oocyst suspension was added to the beaker, which was gently stirred on a magnetic stir plate. Prior to and immediately after oocyst addition, chlorine residuals were determined using the DPD method. Residuals were also determined at 10, 20, 60 and 120 min intervals after oocyst addition.

In order to determine if the 2C3 epitope was chlorine sensitive, intact oocysts were exposed to 4 mg/l chlorine at 22°C, pH 7. The chlorine solution was prepared from a 6% sodium hypochlorite stock solution. Chlorine demand-free buffers and glassware was used. Process control oocysts were subject to same treatment except chlorine was omitted. Treated and control oocysts were reacted at a concentration of 10^6 oocysts per ml with FITC-labeled mAb 2C3 for 1 hr at room temperature. Unbound antibody was removed by precipitating the oocysts. Labeled oocysts were analyzed by flow cytometry using a Coulter EPICS flow cytometer.

CHLORINE TREATMENT

CHAPTER 3
RESULTS AND DISCUSSION

OVERVIEW

The objectives of developing probes such as antibodies for protein components, lectins for carbohydrate components and fluorogenic probes for lipid components were accomplished during the course of this project. The probes were used to study the effect of disinfectants such as chlorine and ozone on these components. These studies are described in detail in the following section. However, cloning of the gene encoding the protein recognized by the mAb 2C3/Cellabs was not successful. In addition, the effects of water treatment processes on components of the oocyst wall could not be studied. Some of these studies are ongoing and will be pursued as part of other projects.

OOCYST ISOLATES AND GENOTYPES

Using different DNA-based techniques, several laboratories (including those at TUSVM), have found that *C. parvum* isolates can be divided into two genetically distinct groups. One, designated genotype I, is almost exclusively associated with human infections; the other, genotype II, causes both human and animal infections (Morgan et al. 1995; Bonnin et al. 1996; Carraway, Tzipori and Widmer 1996; Carraway, Tzipori and Widmer 1997; Morgan et al. 1997; Peng et al. 1997; Spano et al. 1997b; Spano et al. 1998). The existence of two genotypes implies that *C. parvum* may not be a uniform species, rather being comprised of two genetically distinct parasite sub-populations, and suggests the existence of two independent human transmission cycles. In addition to animal infectivity (Widmer et al. 1998), genotypes I and II display phenotypic differences at the oocyst level which might be relevant to the water industry. Although oocysts are known to be stable and survive in water for long periods of time, investigators at TUSVM have observed that some isolates of genotype I are less stable. Whereas the half-lives of bovine and human isolates of genotype II in water at room temperature are around 110 days, some genotype I isolates decay with a half-life of only 6.7 days (Widmer 2002, unpublished data.). These observations raise the possibility that genotypically different *C. parvum* isolates vary in their susceptibility to disinfectants and suggest that a significant degree of heterogeneity exists among genotype I isolates. Most human infections are caused by genotype I isolates and many waterborne outbreaks of cryptosporidiosis in the US have been associated with isolates of this genotype (Fayer, Morgan and Upton 2000). However, most laboratory and water industry-based studies utilize genotype II isolates. To the best of our knowledge prior to this study there have been no reports on biochemical characterization of the oocyst wall of genotype I isolates.

Recently, two genotype I isolates have been identified and characterized at TUSVM. NEMC1 was isolated from an acutely infected individual in January 1999 (Widmer et al. 2000). Since then it has been propagated serially in gnotobiotic piglets at TUSVM. The genotype of oocysts excreted at each passage has been verified by RFLP analysis using 2 markers as described (Spano et al. 1998). The profile of each serial passage was identical to that found in the original human isolate and shows that NEMC1 is of genotype I. At the phenotypic level, oocysts of the pig-passaged NEMC1 isolate were similar to those from some other type I isolates in showing a short oocyst half-life at room temperature (~7 days). This is significantly shorter than the

~110 day half-life of type II oocysts and independently confirms the classification of isolate NEMC1 as type I.

Isolate TU502 was derived from an isolate (UG502) that originated from a Ugandan child with cryptosporidiosis. The UG502 isolate was found to be genetically stable with serial passage in gnotobiotic piglets and was selected for continuous passage in this fashion. UG502 was occasionally propagated in sterilely-delivered calves for large-scale production of the parasite. TU502 was isolated from a human laboratory worker who became accidentally infected with UG502 during this procedure. TU502 and UG502 were found to be genotypically similar by polymerase chain reaction restriction fragment length polymorphism (PCR-RFLP) and microsatellite analysis of multiple loci (Akiyoshi et al. 2002)

In this project the GCH1, UCP, Iowa and MD isolates were used as representatives of genotype II and NEMC1, UG502 or TU502 were used as representative genotype I isolates.

DEVELOPMENT OF PROBES TO IDENTIFY COMPONENTS OF THE OOCYST WALL

Protein Components

Protein Profile of Oocyst Walls

As a first step to characterize oocyst wall protein components the protein profile of oocysts and oocyst walls were examined by one- and two-dimensional SDS PAGE. Aliquots of intact oocysts (GCH1) and Percoll-purified oocyst walls with and without SDS treatment were subjected to SDS PAGE under reducing conditions and proteins identified by silver staining. As seen in Figure 3.1, intact oocysts showed a large number of bands with a wide range of molecular weights, the majority of which probably represent sporozoite proteins. 10 major bands with molecular weights (MW) of ~15, 20, 35, 42, 57, 79, 110, 190 and 2 bands >225 kDa were detected in Percoll-purified oocyst walls and 6 bands with MW of 20, 35, 42, 57, 79 and >225 kDa were detected in SDS-treated walls. By two-dimensional SDS PAGE, a number of spots could be detected in Percoll-purified oocyst walls (not shown) and in walls that had been treated with SDS (Figure 3.2).

Antibody Probes for Specific Oocyst Wall Proteins

In order to determine whether anti-oocyst wall antibodies that were available could serve as probes for study of oocyst wall proteins, the antibodies and antigens with which they react were characterized in some detail.

The murine mAb 2C3 was produced at ICC using oocysts and oocyst walls for initial and booster immunizations respectively. This mAb is of the IgG_3 subclass, and is genus specific. The Cellabs murine IgG mAb to oocyst wall protein/s was obtained from Dr. A.M. Smithyman, Cellabs Pty Ltd, Australia. The rabbit polyclonal antibody anti cpRL3 was generated at ICC and is directed against a polypeptide constituting the amino terminal 786 aa of COWP. The DNA clone encoding this peptide was provided to ICC by Dr. A. Crisanti, Imperial College of Science, Technology and Medicine, London, UK. All three mAbs reacted with oocyst walls (Figure 3.3). The pattern of reactivity of mAb 2C3 and Cellabs mAb appeared to be identical.

Immunoblotting analysis of oocyst walls revealed that mAb 2C3 reacted with bands >225,175 and 79 kDa in MW in oocyst walls (Figure 3.4). mAb 2C3 reacted with intact oocysts as well as with isolated oocyst walls. However, the majority of the antigen recognized by 2C3 was eluted by SDS treatment of oocyst walls, in contrast to that recognized by anti-cpRL3, which remained associated with the SDS-treated walls. There was no difference in mAb 2C3 or Cellabs mAb reactivity with oocysts under reducing or non-reducing conditions (Figure 3.5), whereas anti-cpRL3 reacted with oocysts walls only under reducing conditions (not shown). These results confirm that the 2C3/Cellabs antigen/s is present in oocyst walls and, unlike COWP, do/does not contain inter or intra chain disulfide bonds. Mild periodate treatment had no effect on reactivity with mAb 2C3 (Figure 3.6). This result confirms that the epitope recognized by mAb 2C3 on oocyst walls is not carbohydrate or carbohydrate-dependent.

By IFA, double label experiments demonstrated that 2C3 mAb and anti-cpRL3 polyclonal antibody recognize distinct epitopes (Figure 3.7). While 2C3 recognized the oocyst wall of intact or excysted oocysts, anti-cpRL3 appears to recognize oocyst walls of excysted oocysts preferentially. In addition, there appears a more punctate distribution of fluorescence associated with anti-cpRL3, while mAb 2C3 is very uniform in its distribution. By IEM of infected intestinal tissue, mAb 2C3 showed reactivity with the oocyst wall in sporulated oocysts as well as in unsporulated oocysts and the surrounding parasitophorus membrane (not shown). There was no reactivity with other intracellular forms such as trophozoites or schizonts showing specificity of the antibody for the oocyst wall. In purified oocysts, the colloidal gold label was uniformly present on the oocyst wall, predominantly on the outer layer (Figure 3.8). Taken together the IFA and IEM findings as well as previously published studies on COWP indicate that the 2C3 antigen is present on the outer oocyst wall layer, whereas COWP is an integral component of the inner oocyst wall layer.

Differences Between Genotype I and Genotype II Isolates

One of the objectives of this project was to identify differences, if any, in oocyst wall components between the two main genotypes. In order to determine if there were differences using the two previously chosen antibody probes to the 2C3 antigen and COWP protein, oocysts of both genotypes were electrophoresed on 5-15% SDS gradient gels, transferred to nitrocellulose and probed with mAb 2C3 (culture supernatant diluted 1:00), Cellabs mAb (1 μg/ml) and anti-cpRL3 (1 μg/ml), followed by anti-mouse IgG-HRP conjugate 1:20,000 (for 2C3 and Cellabs) and anti-rabbit-HRP conjugate, 1:20,000 (for anti-cpRL3) respectively. The results with the Cellabs mAb and anti-cpRL3 are shown in Figure 3.9 and indicate that there were significant differences in the molecular weights of protein bands recognized by the Cellabs mAb in genotype I and II oocysts, whereas there were no major differences in those recognized by anti-cpRL3 (similar results were obtained by mAb 2C3). This finding may be consistent with the fact that COWP is an internal, structural oocyst wall protein, whereas the protein/s recognized by mAb2C3/Cellabs are present on the oocyst surface and may therefore be subjected to selective pressure. In other studies we have identified extensive differences in sporozoite surface proteins among isolates of both genotypes.

Attempts to Clone the 2C3/Cellabs Antigen/s

The gene encoding COWP has already been cloned and sequenced. Therefore, cloning of the gene encoding antigens recognized by mAb 2C3 was attempted. Approximately 1×10^7 phages from a λzap II Iowa sporozoite cDNA library (obtained as a gift from Dr. Richard Nelson, UCSF) were screened with supernatant from a 2C3 producing hybridoma. Two plaques potentially positive for the 2C3 antigen were re-screened and determined to be false positives. Sonicated oocysts were used as a positive control antigen preparation to ensure that the antibodies/detection systems worked as expected. All the antibodies detected the controls.

Approximately 5×10^6 phages were then screened with the CrA2 mAb (to gp15, a *C. parvum* surface glycoprotein) as a positive control. Four potentially positive plaques were re-screened and determined to be false positives. An additional positive control antibody, 8A9.2, a mouse α sporozoite IgG antibody developed at TUSVM, was also used to screen the library. This antibody identified several plaques demonstrating that the screening procedure was working as expected. The SuperSignal chemiluminescent detection system gave strong repeatable positive signals with 8A9.2, and alternative detection systems were also explored. The sporozoite cDNA library (2×10^6 plaques) was screened with the 2C3 supernatant using an alkaline phosphatase-conjugated secondary antibody and a NBT/BCIP colorometric kit from Promega. However, no plaques were detected.

Because of the possibility that the 2C3 gene is not expressed in sporozoites, alternative libraries, including a λ zap II library prepared from a partial EcoRI digest of genomic DNA from GCH1 oocysts, two λ zap express libraries prepared from partial Sau3AI digests of genomic DNA from GCH1 and NEMC1 oocysts, a λZAP II GCH1 oocyst genomic library (complete EcoRI digest) and a λgt 11 oocyst (NINC isolate), genomic library, a gift from Dr. Rick Nelson, UCSF) were also screened. Approximately, 1×10^7 phages from each of these libraries were screened with the 2C3 mAb. However, no positive plaques were detected. The *C. parvum* genomic library (complete EcoRI digest) was also screened with the mAb from Cellabs, which recognizes the same antigens as 2C3. No positive plaques were detected in 5×10^6 plaques screened. The specificity of the screening/ development methods was checked by developing dot blots of oocyst proteins along side the library membranes with both 2C3 and the Cellabs mAb. Screening of several *C. parvum* phage libraries with antibody probes failed to identify genes coding for the 2C3 antigen. The lack of detectable 2C3 clones in genomic libraries could be attributed to the absence of the intact coding sequence in the libraries screened, or to the failure of 2C3 expression due to a shifted reading frame in positive clones.

As an alternative approach to using antibody probes to screen libraries in order to clone the gene encoding the antigens recognized by mAb 2C3/Cellabs, an attempt was made to obtain N-terminal and/or internal amino acid sequence from the proteins in order to design degenerate oligonucleotides. However, attempts to obtain N-terminal or internal amino acid sequence were also not successful. For this purpose oocyst wall proteins were separated by one and two-dimensional SDS PAGE, the gels were stained with Problue as well as silver stains or transferred to nitrocellulose or Immobilon and probed with mAb 2C3 or Cellabs mAb. Although proteins in the molecular weight range of those recognized by the mAbs could be visualized with silver staining and by immunoblotting, these bands could not be visualized by Problue staining in gels or blots, indicating that there would not be sufficient protein to obtain sequence. The main problem appeared to be that the oocyst wall proteins recognized by mAb 2C3/Cellabs did not enter the 1 or 2 D gels in sufficient quantity to visualize bands in order to cut them out of gels or blots. This may be because these antigens are heavily glycoyslated, poorly soluble, precipitate or aggregate. Attempts to solubilize the antigens or prevent aggregation using various detergents and buffer for the gels were not successful.

Summary

In summary these experiments suggest that 1) All three antibodies recognize oocyst wall proteins, with mAb 2C3 and Cellabs mAb seemingly recognizing the same epitopes, 2) the cpRL3 epitope contains disulphide-linked subunits whereas the 2C3/Cellabs epitope does not, 3) the 2C3/Cellabs epitope is not carbohydrate or carbohydrate-dependent in nature, and 4) the cpRL3 epitope is present on the inner face of the oocyst, while the 2C3/Cellabs epitope is exposed on the outer face of the oocyst wall. These antibodies are useful tools with which to probe the oocyst wall surface, and to correlate treatments that may change the integrity of the wall with differences in the presence of these molecules. These studies also indicate that the protein recognized by mAb 2C3/Cellabs is different from COWP. This is based on the findings that 1) the molecular weights of the proteins as determined by immunoblotting are different, 2) the 2C3 protein, unlike COWP does not contain inter or intra chain disulfide bonds, and 3) the ultrastructural localization of the 2C3 protein is different from that of COWP. Attempts to clone the gene/s encoding the 2C3/Cellabs antigen/s were not successful.

Carbohydrate Components

In order to explore the possibility of using lectins as probes for specific monosaccharide residues in oocyst wall glycoconjugates, a panel of lectins with varying sugar specificity (Table 3.1) was used to perform fluorescence-based binding assays on fixed oocysts as well as oocysts in suspension.

Lectin Binding to Fixed Oocysts

The results of lectin binding to fixed GCH1 oocysts are shown in Table 3.2. Lectins with specificity for N-acetylgalactosamine (GalNAc), galactose (Gal), fucose, glucose, mannose and N-acetyl glucosamine reacted with oocysts and/or oocyst walls, suggesting the presence of these monosaccharides in oocyst walls. However, the pattern and intensity of reactivity varied with the preparation used and with the lectin. The strongest reactivity occurred with lectins specific for Gal and GalNAc. Examples of three of the most intensely staining of these lectins, namely MPA, HPA, and VVA, are shown in Figures 3.10 to 3.14. Thus, *Maclura pomifera* agglutinin (MPA) which is specific for αGalNAc and αGal residues reacted in a circumferential pattern with untreated as well as with hypochlorite treated oocysts (Figure 3.10). Reactivity was greatly diminished by preincubation of the lectin with GalNAc (Figure 3.10) prior to the assay, confirming the specificity of binding. Isolated oocyst walls as well as SDS-treated oocyst walls also reacted in a circumferential pattern with MPA (Figure 3.11). In oocysts that were not treated with hypochlorite, *Helix pomatia* agglutinin (HPA) which is specific for αGalNAc, reacted predominantly with material that was shed from the oocyst during excystation and with newly excysted sporozoites. VVA reacted in a circumferential pattern with oocysts that were not hypochlorite treated and with shed material in hypochlorite-treated oocysts (Figure 3.14).

Table 3.1
Monosaccharide specificity of lectins used to study oocyst wall carbohydrates and glycoconjugates

Lectin	Monosaccharide specificity
Helix pomatia agglutinin (HPA)	α–N-acetylgalactosamine (GalNAc)
Vicia villosa agglutinin (VVA)	αGalNAc
Artocarpus integrifolia agglutinin (AIA)	αGalactose (Gal), αGalNAc
Maclura pomifera agglutinin (MPA)	αGalNAc, αGal,
Glycine max agglutinin (SBA)	α (β) GalNAc, Gal
Arachis hypogea agglutinin (PNA)	βGal
Sophora japanicum agglutinin (SJA)	βGalNAc, Gal
Ulex europeus agglutinin (UEA-1)	αL-Fucose (L-Fuc)
Canavalia ensiformis agglutinin (Con A)	Glucose (Glc), Mannose (Man)
Triticum vulgare agglutinin (WGA)	N-acetylglucosamine (GlcNAc)
Sambucus nigrans agglutinin (SNA)	N-acetylneuraminic acid (Neu5Ac), βGal
Maakia amurennis agglutinin (MAA)	Neu5Ac

Table 3.2
Lectin binding to *C. parvum* GCH1 fixed oocysts and oocyst walls

Lectin	Specificity	Oocysts	Oocyst walls	SDS-treated oocyst walls
HPA	αGalNAc	++(C)	++(C)	++(C)
AIA	αGalNAc, αGal	++(C)	++(C)	++(C)
MPA	αGalNAc, αGal	+++(C)	+++(C)	+++(C)
VVA	αGalNAc	+++(C)	+++(C)	+++(C)
SBA	α, βGalNAc, Gal	+ (P)	-	-
SJA	βGalNAc, Gal	+(P)	-	-
UEA-1	αL-Fuc	+(P)	+(P)	+(P)
CON A	Glc, Man	++(P)	+(P)	+(P)
WGA	GlcNAc, Neu5NAc	++(P)	+(P)	+(P)
SNA	Neu5NAc, βGal	-	-	-
MAA	Neu5NAc, βGal	-	-	-

See Table 3.1 for abbreviations of lectins and carbohydrates. -, negative; +, weak binding; ++, strong binding; +++, very strong binding; C, circumferential pattern of staining; P, polar pattern of staining

Table 3.3
Lectin binding to *C. parvum* TU502 fixed oocysts and oocyst walls

Lectin	Specificity	Oocysts	Oocyst walls
HPA	αGalNAc	+(P)	++(C)
AIA	αGalNAc, αGal	++(C)	++(C)
MPA	αGalNAc, αGal	+++(C)	+++(C)
VVA	αGalNAc	+(C)	+(C)
PNA	Gal β1–3 GalNAc, Gal	+ (C)	++ (C)
SJA	βGalNAc, Gal	+(P)	-
CON A	Glc, Man	+(P)	+(P)
WGA	GlcNAc, Neu5NAc	+(P)	-(P)
SNA	Neu5NAc, βGal	-	-

See Table 3.1 for abbreviations of lectins and carbohydrates. -, negative; +, weak binding; ++, strong binding; +++, very strong binding; C, circumferential pattern of staining; P, polar pattern of staining

In order to determine if there were any differences in lectin binding between genotypes I and II, we repeated the experiments with TU502 (genotype I) oocysts. In general the results (shown in Table 3.3) were similar to those obtained with GCH1 oocysts (genotype II) except that there was reactivity (particularly of oocyst walls) with PNA which we did not find with GCH1. Examples of lectin binding to TU502 oocysts are shown in Figure 3.15.

Lectin Binding to Intact Oocysts in Suspension

It is possible that fixation of the oocysts may alter conformation or accessibility of the wall carbohydrates. In order to ensure that spatial configurations of the carbohydrates would most closely approximate their biological state lectin binding studies were performed on intact unfixed oocysts in suspension. These studies were performed on both genotype I (TU502) and genotype II isolates (GCH1). Conditions that were somewhat different from those used for fixed oocysts were necessary. For example, all procedures had to be performed at 4°C rather than room temperature, in order to prevent excystation. In spite of efforts to perform all procedures at 4°C some degree (~5-10%) of excystation in the control preparations (buffer alone) was observed. However, this enabled visualization of lectin binding to excysted oocyst walls. In addition, higher concentrations of lectins had to be used to obviate photobleaching, as a mounting buffer could not be used with an anti-photobleaching agent since these studies were performed in suspension. The results of these studies are shown in Tables 3.4 and 3.5. Examples of lectins which reacted positively are shown in Figures 3.16-3.18.

In contrast to the results obtained with fixed oocysts, intact oocysts reacted only with three lectins. The strongest reactivity by fluorescence occurred with MPA (Figures 3.16, 3.17) and AIA (Figure 3.18). Both intact oocysts as well as excysted oocyst walls reacted with these lectins. A striking finding was the marked degree of agglutination that occurred with these two lectins.

This finding can be explained on the basis of the fact that lectins are bi or multivalent and would therefore bind to carbohydrate residues on multiple oocysts causing them to agglutinate.

Table 3.4
Lectin binding to *C. parvum* GCH1 intact oocysts and oocyst walls in suspension

Lectin	Specificity	Oocysts	Oocyst walls	Agglutination	Excystation
HPA	αGalNAc	-	-	-	+
AIA	αGalNAc, αGal	++	+	++	+
MPA	αGalNAc, αGal	+++	++	+++	++
VVA	αGalNAc	+	+	-	+
SBA	α, βGalNAc, Gal	-	-	-	+
SJA	βGalNAc, Gal	-	-	-	+
PNA	βGal, Galβ1–3GalNAc		-	-	+
UEA-1	αL-Fuc	-	-	-	+
CON A	Glc, Man	-	-	-	+
WGA	GlcNAc, Neu5NAc	-	-	-	+
SNA	Neu5NAc, βGal	-	-	-	+
MAA	Neu5NAc, βGal	-	-	-	+

See Table 3.1 for abbreviations of lectins and carbohydrates. -, negative; +, weak ++, strong +++, very strong

Table 3.5
Lectin binding to *C. parvum* TU502 intact oocysts and oocyst walls in suspension

Lectin	Specificity	Oocysts	Oocyst walls	Agglutination	Excystation
HPA	αGalNAc	-	-	-	+
AIA	αGalNAc, αGal	++	+	++	++
MPA	αGalNAc, αGal	++	++	+++	+++
VVA	αGalNAc	+/-	+	-	+
SBA	α, βGalNAc, Gal	-	-	-	+
SJA	βGalNAc, Gal	-	-	-	+
PNA	βGal, Galβ1–3GalNAc	-	-	-	+
UEA-1	αL-Fuc	-	-	-	+
CON A	Glc, Man	-	-	-	+
WGA	GlcNAc, Neu5NAc	-	-	-	+
SNA	Neu5NAc, βGal	-	-	-	+
MAA	Neu5NAc, βGal	-	-	-	+

See Table 3.1 for abbreviations of lectins and carbohydrates. -, negative; +, weak ++, strong +++, very strong

Table 3.6
Effect of peptide-n-glycosidase F treatment on FITC-lectin binding to *C. parvum* oocysts and oocyst walls

Lectin	Specificity	Oocysts (control)	Oocysts (enzyme-treated)	Oocyst walls (control)	Oocyst walls (enzyme treated)
HPA	αGalNAc	++	++	+	+
AIA	αGalNAc, αGal	++	++	+	+
MPA	αGalNAc, αGal	+++	+++	+++	+++
VVA	αGalNAc	++	++	+	+

See Table 3.1 for abbreviations of lectins and carbohydrates. -, negative; +, weak binding; ++, strong binding; +++, very strong binding

Another striking finding was the induction of excystation by both these lectins. This effect was observed to a greater degree with oocysts of the TU502 isolate. The reasons for the excystation are not known at present. The agglutination as well as excystation induced by these two lectins confirms the surface localization of these carbohydrates. The lectin VVA bound weakly to oocysts and did not induce agglutination. There was no reactivity of either isolate with PNA (as found for fixed TU502 oocysts). There were only minor differences in reactivity between oocysts of either genotype. In general TU502 oocysts reacted less strongly by florescence, but appeared to be agglutinated to the same degree. Excystation and consequent agglutination of oocysts by the lectins MPA and AIA occurred more so with TU502 oocysts. The reasons for these differences are not clear.

As described above, using fixed oocysts, in addition to MPA and AIA, it was found that lectins such as HPA, SBA, SJA as well as Con A, WGA and UEA reacted with oocyst walls, albeit less strongly than with MPA and AIA. One of the differences in specificity between these two lectins and the other αGalNAc-specific lectins is the affinity for αgalactosyl residues. It is possible that the outer surface of the oocyst wall contains predominantly αgalactosyl residues, whereas the inner layers contain αGalNAc residues in addition. However, this is speculative and needs to be confirmed by other methods.

Binding of GalNAc-specific lectins to unfixed and fixed oocysts was confirmed by flow cytometry. Results of lectin binding to GCH1 and TU502 oocysts by this method are described in the "effect of ozonation on carbohydrates" section that follows.

Glycosidase Treatment of Oocysts

In order to confirm and extend the results of lectin-binding studies, we next investigated the effect of glycosidase treatment on lectin binding. Initial studies were performed with lectins that reacted most strongly with the oocyst wall, namely αGalNAc-specific lectins, in order to determine the glycosidic linkage of the GalNAc residues. The glycosidases used, their abbreviations and their site of action are shown in Figure 3.19. These glycosidases were chosen to determine whether the GalNAc residues identified by lectin binding were present in N or O glycosidic linkages and to determine whether the GalNAc residues were terminal or internal. The results are shown in Tables 3.6, 3.7, and 3.8.

Table 3.7
Effect of endo-α-n-acetylgalactosaminidase treatment on FITC-lectin binding to *C. parvum* oocysts and oocyst walls

Lectin	Specificity	Oocysts (control)	Oocysts (enzyme-treated)	Oocyst walls (control)	Oocyst walls (enzyme treated)
HPA	αGalNAc	++	++	+	+
AIA	αGalNAc, αGal	++	++	+	+
MPA	αGalNAc, αGal	+++	+++	+++	+++
VVA	αGalNAc	++	++	+	+

See Table 3.1 for abbreviations of lectins and carbohydrates.-, negative; +, weak binding; ++, strong binding; +++, very strong binding

Table 3.8
Effect of α-n-acetylgalactosaminidase treatment on FITC-lectin binding to *C. parvum* oocysts and oocyst walls

Lectin	Specificity	Oocysts (control)	Oocysts (enzyme-treated)	Oocyst walls (control)	Oocyst walls (enzyme treated)
HPA	αGalNAc	++	+	+	-
AIA	αGalNAc, αGal	++	+	+	-
MPA	αGalNAc, αGal	+++	+	++	-
VVA	αGalNAc	++	++	+	-

See Table 3.1 for abbreviations of lectins and carbohydrates. -, negative; +, weak binding; ++, strong binding; +++, very strong binding

PNGase F and Endo-α-GalNAcase had no significant effect on lectin binding indicating that the GalNAc residues were not present in an N-glycosidic linkage and that they were not present as part of the Galβ1-3GalNAc determinant. Treatment with α-GalNAcase, however, resulted in significant reduction of binding of all the GalNAc-specific lectins, (results with MPA are shown in Figure 3.20) suggesting that these residues are terminal. This pattern of lectin binding before and after treatment with α-GalNAcase is consistent with the presence of the Tn (GalNAc α1-Ser/Thr) antigen, which is the monosaccharide precursor of the O-linked core disaccharide, T (Thomsen Friedenreich) antigen. These determinants are characteristic of mucin-type O-glycosylation. These results suggest that this type of glycosylation may be present in oocyst wall glycoproteins and/or glycolipids.

Monosaccharide Analysis of Oocyst Walls

In order to confirm the presence of monosaccharides suggested by the lectin binding studies, 10^8 Percoll-isolated oocyst walls (GCH1) were sent to the UCSD Glycotechnology Center Core facility in consultation with Dr. Herman van Halbeek for monosaccharide analysis by HPAEC PAD. The sample was subjected to acid hydrolysis, followed by HPAEC-PAD of the resulting mixture of monosaccharides and other components. The HPAEC profile (the result of injection of 10% of the total hydrolysate), along with the profile of a standard monosaccharide mixture are shown in Figure 3.21. The profile of the oocyst-wall hydrolysate shows large amounts of glucose (Glc), smaller amounts of galactosamine (presumablyGalNAc before hydrolysis), galactose (Gal), mannose (Man), and glucosamine (GlcNAc before hydrolysis). Qualitatively, this result is in agreement with our lectin-binding studies. However, it was difficult to quantitate the molar ratios of other monosaccharides in the presence of the large excess of Glc. Previous studies (Nanduri et al. 1999) have also documented large amounts of glucose in oocyst walls. Current efforts are directed at confirming these results and also at excluding the possibility of contamination of the oocyst walls with Glc during the oocyst purification process.

Summary

In summary, a panel of lectins of varying specificity were used as probes to investigate the nature of carbohydrate residues present in the oocyst wall. Binding of FITC-conjugated lectins to intact oocysts of both genotype I and II, oocyst walls and SDS or hypochlorite treated oocyst walls was qualitatively assessed by microscopy. Reactivity of the lectins with fixed oocysts as well as intact oocysts in suspension was determined. The results indicated that lectins with specificity for GalNAc, galactose Gal, fucose, glucose, mannose and N-acetyl glucosamine reacted with oocysts and/or oocyst walls, suggesting the presence of these monosaccharides in oocyst walls. However, the pattern and intensity of reactivity varied with the conditions used and with the lectin. The strongest reactivity occurred with lectins specific for Gal and GalNAc. Binding of Gal/GalNAc-specific lectins HPA and MPA was confirmed and quantitated by flow cytometry. In contrast to the results obtained with fixed oocysts, intact oocysts reacted only with three lectins. The strongest reactivity by fluorescence occurred with MPA and AIA. A striking finding was that these lectins agglutinated oocysts and also induced excystation of intact oocysts in suspension.

Glycosidases were also used as probes to confirm the lectin binding findings and to determine the type of glycosylation present. PNGase F and Endo-α-GalNAcase had no significant effect on lectin binding indicating that the GalNAc residues were not present in an N-glycosidic linkage and that they were not present as part of the Galβ1-3GalNAc determinant. Treatment with α-GalNAcase however resulted in significant reduction of binding of all the GalNAc-specific lectins, suggesting that these residues are terminal. The presence of large amounts of Glc, smaller amounts of galactosamine (presumably GalNAc before hydrolysis), Gal Man, and glucosamine were qualitatively confirmed by HPAEC.

Table 3.9
Fluorogenic lipid probes used

Probe #	Abbreviation	Lipid class	Ex/Em
1	NBD-C_6-HPCl	Phospholipids	470/530 nm
2	NBD-C_{12}-DPC	Phospholipids	470/530 nm
3	NBD-C_6-sphingomyelin	Sphingolipids	470/530 nm
4	NBD-Sterol	Cholesterol	468/538 nm
5	AOA	Fatty Acid	360/445 nm

1. 2-(Nitrobenzoxadiazolyl-amino)hexanoyl-1-hexadecanoyl-sn-glycero-3-phosphocholine
2. 2-(Nitrobenzoxadiazolyl-amino)dodecanoyl-1-hexadecanoyl-sn-glycero-3-phosphocholine
3. 6-((N-7-Nitrobenz-2-oxa-1,3,diazol-4-yl)amino)hexanolyl)sphingo-yl phosphocholine
4. 22-(*N*-(7-nitrobenz-2-oxa-1, 3-diazol-4-yl)amino)-23,24- bisnor-5-cholen-3β-ol
5. 12-(9-anthroyloxy)oleic acid

Table 3.10
Staining conditions for fluorogenic lipid probes

Condition/treatment	Probe used	Result
MDBK cells (+ve control)	All	Characteristic staining of eukaryotic lipids
Intact oocysts	All	No staining (NS)
Excysted oocysts	All	Staining of sporozoites
Isolated oocyst walls	All	NS
Freeze/thawed oocysts	All	NS (intact oocysts or shells)
Hypochlorite	All	Slight staining with all probes
1% SDS/boil	All	Fluorescence of 'inner structure'

Lipid Components

Fluorogenic Lipid Probe Binding to Oocysts

The objective of this part of the project was to characterize the presence of various classes of lipids in oocyst walls using fluorogenic probes. The probes, their characteristics and the staining conditions and treatments are listed in Tables 3.9 and 3.10, respectively. All five of the probes were generally unsuccessful in staining intact oocysts. Less than 1% of the oocysts fluoresced intensely enough to be observed. Most oocysts were either very dimly illuminated or unobservable under fluorescence.

In order to confirm that the probes were functional, cultures of Madin-Darby Bovine Kidney cells (MDBK) prepared in Lab-Tek® Chamber Slides were analyzed for reactivity with the probes. All five of the probes fluoresced brightly with MDBK cells, indicating the presence of the standard lipid classes known to be present in eukaryotic cells. Interestingly, each probe produced a characteristic staining pattern (with differential staining of plasma membrane and resident organelles such as golgi, nuclear envelope, etc.), probably reflecting unique lipid composition of those organelles.

Excystation and oocyst wall isolation was attempted to determine whether wall isolates would take up the probes more successfully than the intact oocysts. Preliminary results indicated that the lipid probes failed to stain isolated oocyst walls. This may indicate an inner layer that is impermeable to the probes, or the collapsed nature of the isolated walls precluded probe entry (this latter explanation is less likely, given that antibody probes can enter and stain interior wall structures).

The next series of experiments was designed to compromise this structural integrity and attempt staining. The first of such attempts was freeze-thaw treatment. Oocysts were frozen at –80°C, followed by thawing. Under microscopic examination the oocysts appeared to be damaged and the sample had high concentrations of debris (consisting of unknown particles and sporozoites). Under fluorescence what intact oocysts remained were still non-fluorescent.

Preliminary testing of hypochlorite exposure on the oocyst showed very little variance from stains of whole-untreated oocysts. This procedure was performed on both UCP and GCH1 oocysts. Whole oocysts seemed to be more noticeable under fluorescence after hypochlorite treatment. Oocysts still fluoresced dimly, but of 100 oocysts counted ~95% were bright enough to observe. This may support the concept of hypochlorite 'shaving' of outer wall layers, eventually making the oocyst more permeable to small molecules.

Oocysts were then treated with SDS to potentially disrupt any surface protein structure associated with the oocyst wall. Using phase microscopy the oocysts seemed to be intact but an inner membrane-like structure had retracted from the outer wall surface. This inner body refracted the light and it appeared to be disconnected completely from the oocyst wall (Figure 3.22). All the oocysts fluoresced using all five probes, when treated with SDS in this manner. The inner refractive piece fluoresced brightly, while the outer wall fluoresced less intensely. The five probes produced the same staining pattern. Results using NBD-C_{12}-HPC are shown in Figure 3.23. Oocysts treated in SDS were used as standards to which the success of lipid extractions was measured. It is possible that the structures visualized in this manner were the agglutinated membranes of sporozoites after the SDS lysis step.

Solvent Extraction

The procedures and results of solvent extractions are summarized in Table 3.11. Oocysts treated with ethanol after SDS were clumped and distorted. The oocysts fluoresced; however it was difficult to determine whether the extraction had caused a difference in the SDS-treated staining pattern. The few lone, non-clumped oocysts retained the same pattern as those in the control sample.

Hexane extraction after SDS treatment did not affect the staining pattern. Chloroform:methanol extraction after SDS treatment damaged the oocysts and many large clumps of debris were observed. The coagulated oocyst masses fluoresced brightly but the intensity of the fluorescence was likely due to the trapping of the flurogenic probes in these clumps. These aggregates appeared at the solvent:aqueous interface, so the results were interpreted as non-specific entrapment of stain in the denatured proteinaceous mass. There were discernible oocysts in these clumps, and they exhibited the 'SDS' staining pattern. Approximately 5% of the oocysts in suspension remained intact, exhibiting the typical SDS pattern.

Table 3.11
Effect of solvent extractions on binding of fluorogenic lipid probes

Treatment	Desired effect	Result
SDS/boil, hexane, stain SDS/boil	Remove non-polar lipids	No effect on fluorescence
$CHCl_3$: CH_3OH (2:1), stain	Remove non-polar lipids	Massive denaturation/agg
SDS/boil, ethanol, stain	Remove polar lipids	No effect on fluorescence
Acetone, SDS/boil, stain	Remove polar lipids	Oocyst shrinkage, staining
Hexane, SDS/boil, stain	Remove non-polar lipids	No effect on staining
$CHCl_3$: CH_3OH (2:1) SDS/boil, stain	Remove non-polar lipids	Significant losses of oocysts with fluorescing inner structure
0.4 M NaAcetate CH_3OH: $CHCl_3$:H_2O (60:30:8), SDS/boil, stain	Remove sphingolipids	No effect on staining
Ethanol, SDS/boil, stain	Remove polar lipids	Oocyst shrinkage, staining

Extractions were attempted prior to SDS treatment, with the notion that microscopic analysis would be more successful due to less aggregation and generation of fluorescing debris. Extraction with acetone caused the oocysts to become smaller, losing about half a diameter length in size. It was inconclusive whether the SDS pattern remained because the oocysts were too small to observe the presence of an inner membrane. While hexane extractions produced the same results as observed with oocysts extracted after SDS treatment, chloroform:methanol extractions exhibited a different result. Many of the lone oocysts lost the inner structure generated by SDS treatment. It is conceivable that chloroform:methanol extraction was successful in extracting non-polar lipids from intact oocysts, thus rendering them lipid-poor and not stainable after SDS treatment. Alternatively, sporozoite membranes had lost their lipids and the inner structure could not be formed by SDS treatment.

Extractions with 0.4 M sodium acetate methanol:chloroform:water solution were conducted before SDS treatment, and the majority of the oocysts retained the SDS pattern seen in the control samples. This indicates that either sphingolipids are not a significant component of this inner structure, or the solvent did not gain entry into the oocyst interior. Ethanol extraction was repeated causing a similar dehydration effect observed in oocysts treated with acetone, showing the same fluorescent profiles.

Fluorogenic Lipid Probe Binding to Isolated Oocyst Walls

In the next set of experiments intact oocysts (GCH1 and TU502) as well as isolated oocyst walls were analyzed for lipid probe binding. The objective of these experiments was 1) determine if the inner membrane-containing structure that fluoresces in whole oocysts when treated with SDS and heat is denatured sporozoites within the oocyst, and 2) classify lipid species in oocyst walls of *C. parvum* using fluorogenic lipid probes. Intact and SDS + heat-treated whole oocysts were stained for phospholipids (NBD-C_6-HPC) and cholesterol derivatives (NBD-Sterol). In addition, a control was implemented in which non- and SDS + heat-treated, isolated oocyst walls were stained for phospholipids (NBD-C_6-HPC), sphingolipids (NBD-C_6-sphingomyelin), and cholesterol derivatives (NBD-Sterol).

Though non-treated, isolated oocyst walls did not incorporate NBD-C_6-HPC staining, SDS and heat-treated oocyst walls brightly stained with NBD-C_6-HPC or NBD-C_6-sphingomyelin probes (Figures 3.24-3.28). Interestingly, as illustrated in Figure 3.27, SDS and heat-treated oocyst walls failed to incorporate fluorescent cholesterol probes. A common feature observed using DIC optics was that oocyst walls treated with SDS and heat appeared to be delaminated; that is, regions of the oocyst wall had separated from the intact structure seen in non-treated, isolated oocyst walls (Figures 3.24-28). These regions also fluoresced with fluorescent NBD-C_6-HPC and NBD-C_6-sphingomyelin probes (Figures 3.24-3.25).

As in non-treated, isolated oocyst walls, non-treated whole oocysts did not incorporate NBD-C_6-HPC staining (Figure 3.28). On the other hand, 90% or more whole oocysts fluoresced with NBD-C_6-HPC after being treated with SDS and heat. Fluorescing oocysts after SDS and heat treatment and NBD-C_6-HPC staining are shown in Figure 3.29. Both the oocyst walls and internal membrane-containing structure fluoresced (Figure 3.29). Moreover, it was clear that SDS and heat-treated whole oocysts that were stained with NBD-Sterol expressed fluorescent internal membrane-containing structures but not fluorescent oocyst walls (Figure 3.30). In addition, as seen in isolated oocyst walls using DIC optics, regions of the oocyst walls were retracted in the whole oocysts (Figures 3.28-3.30). These regions fluoresced when treated with NBD-C_6-HPC, but not with NBD-Sterol, which is consistent with the isolated oocyst wall controls (Figures 3.25, 3.26, 3.27, 3.29, 3.30).

An interesting finding was that a greater number of TU502 oocysts fluoresced than GCH1 oocysts did, which may be consistent with thicker or more resistant external layers in GCH1 oocysts in comparison to that in TU502 oocysts. Applying fluorogenic lipid probes to oocysts at temperatures above the melting point of long chain 'waxes' was also not successful in permitting lipid staining. It was thought that at elevated temperatures fluidity of the oocyst wall would increase, thus making fluorogenic lipid probes more soluble. Indeed long chain hydrocarbons have been reported in oocyst wall preparations (Anthony et al. 1997).

SDS and heat treatment appear to denature proteins on the outer layer of the oocyst wall, thus allowing fluorogenic lipid probes to bind to structural components of the oocysts. No fluorescent lipid probes were incorporated in the oocyst walls or 'internal structures' unless treated with SDS and heat, which disrupts protein structures associated with the oocyst wall. By doing so, SDS and heat treatment delaminates the oocyst walls, thus allowing the fluorescent lipid probes to integrate with the classes of lipids present in the layers of the oocyst walls. In addition, the resulting delaminated oocyst walls allow for the fluorescent lipid probes to pass across the oocyst walls and into the internal structures that contain various classes of lipids including: phospholipids, sphingolipids, and cholesterol.

It does not appear that the inner membrane-containing structure in whole oocysts that fluoresces for phospholipids, sphingolipids, cholesterol, and fatty acids is an inner layer of the oocyst wall that retracts and completely disconnects from the outer layer of the oocyst wall. Instead, it appears that this inner membrane-containing structure is actually collapsed sporozoite membranes. One piece of supporting evidence is that phospholipids, sphingolipids, cholesterol, and fatty acids are all present in sporozoite membranes while oocyst walls lack cholesterol. Another piece of evidence is that isolated, SDS + heat-treated oocyst walls do not have any retracted and completely disconnected inner layer that fluoresces, regardless of whichever lipid probe is used.

However, it is clear that certain classes of lipids are present in the oocyst walls. Staining of oocyst walls that are isolated or intact in whole oocysts that have been treated with SDS and heat indicate that phospholipids and sphingolipids are present in the oocyst walls whereas cholesterol is not. Moreover, it appears that the oocyst walls stain more brightly with NBD-C_6-sphingomyelin than with NBD-C_6-HPC, which suggests that the oocyst walls contain a higher ratio of sphingolipids than phospholipids.

Summary

In summary, fluorogenic lipid probes were used to analyze the classes of lipids present in oocyst walls. The results of these studies showed that intact oocysts and isolated oocyst walls failed to stain with the fluorogenic lipid probes tested. However, though non-treated, isolated oocyst walls did not incorporate NBD-C_6-HPC staining, SDS and heat-treated oocyst walls brightly stained with NBD-C_6-HPC or NBD-C_6-sphingomyelin probes. Staining of oocyst walls that have been treated with SDS and heat indicate that phospholipids and sphingolipids are present in the oocyst walls whereas cholesterol is not. Moreover, it appears that the oocyst walls stain more brightly with NBD-C_6-sphingomyelin than with NBD-C_6-HPC, which suggests that the oocyst walls contain a higher ratio of sphingolipids than phospholipids. These data are consistent with a model of a tough outer layer present in *Cryptosporidium parvum* oocyst walls that must be denatured in order to incorporate any fluorescent lipid probes. For the three lipid classes examined, the following hierarchy appears to hold for the lipids of *C. parvum* oocyst walls: sphingolipids > phospholipids >> cholesterol.

EFFECT OF DISINFECTANTS ON COMPONENTS OF THE OOCYST WALL

Protein Components

Effect of Chlorination on Proteins

In order to determine if the 2C3 epitope was chlorine sensitive, intact oocysts (genotype II) were exposed to 4 mg/l chlorine at 22°C, pH 7. As seen in Figure 3.31, there was a reduction in fluorescence with increasing exposure time. Chlorine-treated oocysts (10^6/lane) were also subjected to 5-15% SDS PAGE under reducing conditions, and stained with Problue (to control for the amount of protein loaded) or transferred to nitrocellulose and probed with mAb 2C3. A seen in Figure 3.32, there was a reduction in reactivity with increasing time of exposure to chlorine. These results confirm that the epitope recognized by mAb 2C3 is chlorine-sensitive. It is possible that this treatment results in break down of proteins.

Effect of Ozonation on Proteins

Experiments were performed to assess whether the antigen recognized by 2C3 antibody is affected by ozone treatment. To this aim, *C. parvum* oocysts of type I and type II were exposed to 1.4 mg/l ozone at pH 7 and 22°C. GCH1 (genotype II) and TU502 (genotype I) oocysts were exposed for 1, 5, and 10 min. Treated and control oocysts were then reacted with 2C3 antibody.

2C3 binding was visualized by labeling with a secondary FITC labeled antibody. FITC fluorescence was quantitated by flow cytometry. Post-treatment infectivity was assessed in parallel in cell culture. These experiments were repeated once with oocysts from the same lot to assess the reproducibility of the results. As shown in Figure 3.33, 2C3 fluorescence was not affected by ozone. The absence of any ozone effect on 2C3 reactivity contrasts with the loss of infectivity of the oocysts in culture (Figure 3.34). Ozone-treated GCH1 as well as TU502 oocysts (10^6/lane) were also subjected to 5-15% SDS PAGE under reducing conditions, transferred to nitrocellulose and probed with mAb Cellabs and anti-cpRL3. As seen in Figures 3.35 and 3.36, there was no effect on reactivity of either antibody with oocysts of either genotype with increasing time of exposure to ozone. These results confirm that the epitope recognized by both antibodies is ozone-insensitive.

Carbohydrate Components

Effect of Ozonation on Carbohydrates

To quantitate the effect of ozone exposure on oocyst carbohydrates, fresh and paraformaldehyde fixed oocysts were labeled with FITC-conjugated lectins. Paraformaldehyde fixation was necessary to avoid the induction of excystation observed when oocysts were exposed to HPA-FITC. Paraformaldehyde fixation followed by methanol permeabilization was chosen on the basis of previous observations showing intense oocyst labeling with MPA-FITC.

Effect on MPA Binding. Figures 3.37-3.38 show MPA-FITC fluorescence histograms obtained from GCH1 and TU502 oocysts exposed to ozone. The experiment was carried out in duplicate because of the unusually high ozone demand during the first exposure of GCH1 oocysts. Whereas TU502 fluorescence was marginally reduced by ozone (Figure 3.38), no such effect was detected with GCH1 oocysts in the first experiment (Figure 3.37), possibly reflecting the high ozone demand of the oocyst suspension. Alternatively, intrinsic differences in the effect of ozone on the oocyst wall of type I and type II oocysts could account for the differential lectin labeling, perhaps indicating that carbohydrates are more accessible in type I oocysts following ozone exposure. However, because of the relative small differences in fluorescence intensity between treatments and controls, and the differences in fluorescence between process and trip controls, these conclusions are tentative.

In order to discriminate between these 2 possibilities, oocysts of both types were used for a second ozone exposure (Figures 3.39, 3.40). Oocysts from the second exposure experiment were analyzed fresh, i.e., without prior fixation or permeabilization, to assess the effect of ozone on carbohydrates on the native oocyst wall. Control experiments with both types of oocysts showed that lectin binding was specific. The FITC signal obtained with MPA which had been pre-incubated with GalNAc was approximately 7-fold smaller than with the non-competed lectin (not shown) indicating the specificity of the MPA binding reaction. Figures 3.39 and 3.40 illustrate the flow cytometry analysis of both types of oocysts reacted with MPA-FITC. As seen in the first experiment, the effect of ozone on lectin binding was minor, with both types of oocysts. FITC fluorescence was slightly reduced as compared to the process control in all treated oocysts. This effect did not appear to be dose dependent. However, particularly with TU502, we noticed a difference in fluorescence between trip and process control, suggesting that other variables, in addition to ozone affect lectin binding.

Effect on HPA Binding. Ozonation of fixed and permeabilized GCH1 and TU502 oocysts had a visible effect on HPA binding. As illustrated in Figures 3.41 and 3.42, this effect was opposite when the two types of oocysts were compared. GCH1 oocysts exposed to ozone displayed weaker fluorescence than the controls (Figure 3.41). In contrast, TU502 oocyst exposed to ozone fluoresced more brightly than the controls (Figure 3.42). The ozone effect was not dependent on exposure time. Unexposed oocysts of both types displayed similar bimodal fluorescence histograms. Scatter vs FITC fluorescence dot plots indicate that essentially all fluorescent signals originated from oocysts (not shown), consistent with the assumption that 2 populations of oocysts with different lectin affinity were present in the trip control samples. Control histograms from GalNAc inhibited lectin labeled oocysts, and mock-labeled oocysts showed only background level fluorescence (not shown).

In conclusion, FCM analysis of genotype I and genotype II oocysts revealed differences in the effect of ozone on the binding affinity of HPA and MPA. Whereas MPA binding was only marginally or not affected by ozone, HPA binding was either increased (TU502) or decreased (GCH1) in response to ozone. MPA binding could only be tested with fixed oocysts as this lectin was found to trigger excystation.

Effect of Chlorination on Carbohydrates

Using a similar approach as described in the previous section, the effect of chlorination (4 mg/l, pH 5, 22°C) on oocyst lectin binding was investigated. Oocysts were exposed to chlorine at pH 5 to increase the concentration of hypochlorous acid, thereby maximizing the effect of the treatment on the oocysts since pH 5 is the lowest pH at which essentially all chlorine is present in its active form. However, these pH conditions were not intended to mimic water treatment conditions.

Structural Analysis of Chlorine-Treated Oocysts. An ultrastructural analysis of chlorine-treated oocysts was performed using scanning electron microscopy to assess the impact of chlorination on the structural integrity of the oocysts. Although not specifically a part of the scope of this project, we took advantage of excess treated and control oocysts to repeat the electron microscopic analyses as performed previously with *C. parvum* (AwwaRF project 492) and *G. lamblia* cysts exposed to ozone (Widmer et al. 2002). Oocyst samples taken after 300 min, 600 min, and 1200 min were processed for electron microscopy together with two process control samples, sodium thiosulphate quenched and unquenched (Figure 3.43). Confirming previous work, SEM showed no structural differences to the oocyst wall which were associated with treatment.

Carbohydrate Analysis of Chlorine-Treated Oocysts. Chlorine treated and control oocysts were labeled with FITC-MPA and FITC-HPA and analyzed by flow cytometry essentially as described above for the ozone experiments. The goal of this analysis was to assess the effect of chlorine on the integrity of oocyst wall carbohydrates and compare any effect of the chlorine disinfectant with that induced by ozone. We were unable to perform a parallel analysis with type I oocysts, as oocysts of this type were unavailable at the time. Figure 3.44 shows the fluorescence histograms of fixed FITC-MPA labeled GCH1 oocysts. We observed a chlorine dose-dependent increase in oocysts fluorescence, but this effect was relatively small compared to the ~100-fold difference in fluorescence observed between unlabeled oocysts and oocysts labeled in the presence of 0.9 M GalNAc competitor. A similar analysis was performed with unfixed GCH1 oocysts. Only minor differences between the fluorescence histograms of oocysts exposed to chlorine for different time periods were observed (not shown).

Fixed GCH1 oocysts were also reacted with FITC-labeled HPA. A slight increase in fluorescence was detected with increasing exposure to chlorine (Figure 3.45). There was a minor increase in fluorescence which appeared to correlate with the time of exposure to chlorine, but, as seen with MPA, the signal-to-noise ratio was relatively large. There was little increase in fluorescence over the GalNAc-inhibited labeling reaction. The histograms were reminiscent of the results obtained with ozone treated oocysts (Figure 3.41). As in the ozone experiment two populations were identified in the process control samples. Only the low-fluorescent oocysts were detected following treatment.

Based on these observations, the effect on lectin binding of different conditions encountered by mock treated oocysts were examined. Significantly, these treatments resulted in differences in oocyst fluorescence profiles (Figure 3.46), which were also observed with inactivated oocysts. The conclusion from this experiment is that lectin binding is not only affected by the exposure to disinfectants, but also by oocyst storage condition.

Control Experiments Using Mock-Treated Oocysts. The comparison of untreated control oocysts, trip control oocysts and process control oocysts in flow cytometry analyses consistently revealed differences in fluorescence, which was unrelated to the disinfectant. Since the difference in fluorescence among these control oocysts in some cases exceeded the differences observed after exposure to chlorine, the origin of this variability was investigated. Without a better understanding of what affects the binding of lectin to the oocysts, the interpretation of the lectin fluorescence remains tentative. In this experiment unfixed GCH1 oocysts were subjected to three treatments, which were designed to reproduce the conditions that oocysts experience during shipment and during mock exposure (process controls). The following treatments were applied: 1) "trip control": oocysts stored for 20 hr at 4°C in water; 2) "room temperature control," oocysts stored in sodium phosphate buffer at room temperature, and 3) "process control," oocysts stored in sodium phosphate buffer at 4°C. For each "treatment," oocysts were left unlabeled, were labeled with FITC-MPA or were labeled in the presence of 0.9 M GalNAc. This experiment was performed in duplicate using fresh and fixed oocysts. All samples were labeled simultaneously with FITC-MPA. The fluorescence histograms obtained with the fresh oocysts are shown in Figure 3.46. Although background fluorescence (unlabeled oocysts) was not affected by the treatments (Figure 3.46, left column), significant differences between treatments were apparent for the labeled and labeled plus GalNAc samples. These results demonstrate that lectin binding is significantly affected by oocyst storage conditions, and that this effect may exceed any effect of the disinfectant on lectin binding. Differences in the fluorescence histograms were also observed with paraformaldehyde fixed oocysts (not shown).

In spite of the variability among the fluorescence histograms from different control oocyst samples, we noticed that oocysts labeled with FITC-MPA in the presence of GalNAc consistently showed two fluorescence peaks, differing by about 1.5-fold in fluorescence intensity. In order to investigate the origin of these two peaks, the individual peaks were sorted by FACS at the Tufts-New England Medical Center core facility. The individual populations recovered in this manner were analyzed by transmission electron microscopy and the stability of the oocysts in water was measured as previously described (Widmer et al. 2000). None of these analyses revealed any differences between the two oocyst populations. It was our original hypothesis that the two peaks might represent thick and thin-walled oocysts, which are thought to be produced during an infection and have been postulated to be formed during the parasite's life cycle. Electron micrographs obtained from the sorted oocysts showed no structural difference and the oocysts displayed typical stability in water, decaying to 80-85% of the initial count over a 19-day incubation period in sterile water at room temperature. At this point the cause for the differential lectin binding by these oocyst populations is not known.

Lipid Components

Effect of Ozonation on Lipids

The purpose of these experiments was to assess whether ozone affects the uptake of fluorescent lipid probes into oocysts. There was also an investigation of whether oocysts of type I and type II respond differently to ozone with respect to the uptake of the fluorescent phospholipid NBD-C_6-HPC.

The effect of ozone treatment on uptake of fluorogenic lipid probes was initially assessed by microscopy. The results indicated that in comparison to trip controls and process controls, ozone-treated GCH1 and TU502 oocysts did not fluoresce more prominently with NBD-C_6-HPC.

Quantitative measurements of dye uptake were performed by flow cytometry. Microscopic observations (see previous “fluorogenic lipid probe binding to oocysts” section) which showed that boiling of oocysts in 1% SDS for 10 min results in absorption of more NBD-C_6-HPC, resulting in higher fluorescence was confirmed. Exposure to 1.4 mg/l ozone at pH 7 and room temperature also resulted in a small to moderate increase in dye uptake, both in boiled and native oocysts. The response of genotype I and genotype type II oocysts to ozone was similar.

A pilot experiment was performed to assess the effect of SDS/boiling on the uptake of NBD-C_6-HPC (Figure 3.47). Confirming microscopic observations, boiling enhanced the uptake of NBD-C_6-HPC. Fluorescent histograms obtained by flow cytometry also showed uptake of dye into SDS treated, unboiled oocysts and the appearance of 2 differentially stained oocysts populations in unboiled and boiled oocysts. The origin of these multiple peaks was not further investigated.

Ozone Experiment 1. Figure 3.48 summarizes the results with oocysts of genotype II isolate MD exposed to ozone. Fluorescent phosphocholine uptake into boiled oocysts was affected by the disinfectant. Oocysts exposed to ozone for 5 min and 10 min showed the highest fluorescence intensity. Secondary peaks fluorescing more intensely were also observed in these samples. In view of previous observations that sodium thiosulphate added as a quenching agent to process control cysts of *G. lamblia* can affect cyst morphology and biochemical composition (Widmer et al. 2002), two process controls, one with, the other without sodium thiosulphate, were included. As shown in Figure 3.48, process control oocysts emitted approximately 40 times more fluorescence than trip control oocysts, suggesting incomplete labeling of the trip control sample. Dye uptake by unboiled oocysts was, as expected from the preliminary experiment, significantly lower than that of the boiled oocysts. The magnitude of the fluorescent signal was 2-4 times above the fluorescence background emitted by unstained control oocysts. Ozone treatment increased the uptake of NBD-C_6-HPC into unboiled oocysts by about 2-fold. A minor peak fluorescing more intensely was present following a 10 min exposure.

Ozone Experiment 2. A second experiment was performed to assess the reproducibility of the results from the first ozonation experiment, and to compare the response to ozone of genotype I and genotype II oocysts with respect to the uptake of NBD-C_6-HPC (Figure 3.49). As observed in experiment 1, exposed oocysts showed the brightest fluorescence. In contrast to the first experiment, boiled trip control oocysts fluoresced more brightly, and the signal emitted was >1000 times that emitted by unstained oocysts, consistent with the assumption that this sample was incompletely labeled in the first experiment. Results obtained with unboiled oocysts were similar as seen in the first experiment, and the oocysts exposed for 5 min and 10 min displayed higher fluorescence than control oocysts. Again, a minor peak of brightly fluorescing oocysts was observed in the two treated samples and also in the quenched process control. The fact that the oocysts used for the second exposure were 1 month older than those used for the first experiment did not appear to affect the results.

The response of TU502 oocysts (genotype I) to ozone (Figure 3.50) was comparable to that of MD oocysts. Again, 5 min and 10 min exposures to NBD-C_6-HPC resulted in oocysts with the highest fluorescence. Process control oocysts fluoresced more brightly than trip controls, although the difference was not as pronounced as seen with MD oocysts in the first experiment. Unboiled oocyst samples subjected to 10 min ozonation displayed an intensely fluorescent subpopulation.

Effect of Chlorination on Lipids

Effect of Chlorine Experiment 1. To assess the effect of exposure to chlorine on oocyst uptake of labeled phospholipid, oocysts were exposed to 4 mg/l chlorine (pH 5, 22°C) and labeled with NBD-C_6-HPC. As in previous experiments, oocysts were exposed to chlorine at pH 5 to increase the concentration of hypochlorous acid, thereby maximizing the effect of the treatment on the oocysts, since pH 5 is the lowest pH at which essentially all chlorine is present in its active form. Again, these pH conditions were not intended to mimic water treatment conditions.

As in previous experiments, prior to labeling, oocysts were either boiled in the presence of 1% SDS or left unboiled. Chlorine exposure increased uptake of NBD-C_6-HPC (Figure 3.51) into unboiled oocysts, but no effect was seen with the corresponding boiled oocysts, raising the possibility that the presence of a high concentration of fluorescent dye in the boiled samples may lead to saturation or to fluorescence quenching. Experiments in which oocysts were labeled with serially diluted NBD-C_6-HPC dye supported the possibility of concentration dependent quenching.

Fluorescence histograms of labeled unboiled oocysts (Figure 3.51, right column) revealed the presence of two oocyst populations, one displaying background fluorescence, and a second population displaying about 20 times higher fluorescence. The proportion of oocysts in the two populations was not dose dependent, nor was the intensity of the fluorescent signal. Although the origin of these two peaks was not further investigated, the highly fluorescent peak could represent inactivated oocysts, which were made permeable by chlorine.

Effect of Temperature and Buffer on Oocyst Dye Uptake. In an attempt to understand the impact of manipulations performed during disinfection on oocyst labeling with NBD-C_6-HPC, two experiments were performed to mimic the conditions experienced by process and trip control oocysts (temperature, centrifugation, buffer). Oocysts were then boiled, or left unboiled, and the amount of NBD-C_6-HPC uptake measured by flow cytometry. In the first experiment, oocysts were incubated for 2 hr at room temperature or at 4°C, subjected to centrifugation (7000 × g, 10 min) or not centrifuged. Half of the oocysts from each treatment was boiled in SDS and all samples labeled with 1 mg/ml NBD-C_6-HPC. The experiment (Figure 3.52) showed no effect of temperature and centrifugation on dye uptake. Differences in fluorescence between boiled and unboiled oocysts were about 100-fold, but no differences between treatments were visible.

In the second experiment (Figure 3.53), oocysts were incubated in phosphate buffer and water to mimic the buffer conditions of the process and trip control, respectively. A <2-fold increase in fluorescence was observed with the boiled samples, where dye uptake was slightly increased following incubation and stirring in phosphate buffer. The duration of this treatment did not appear to affect the magnitude of the increase in fluorescence. This observation is consistent with results from previous experiments (Figures 3.49-3.51).

Dye Dilution Experiment. To address the possibility that dye saturation or concentration quenching may reduce the ability of detecting differences among samples, an experiment was performed in which oocysts were labeled with serial dilutions of NBD-C_6-HPC. Boiled oocysts were labeled with the same concentration of NBD-C_6-HPC used previously (1 mg/ml), or a 10-, 50-, or 100-fold dilution of the same dye. As shown in Figure 3.54, dye concentration did not significantly reduce the fluorescence. However, at lower dye concentrations, two oocyst populations differing by about 60% in fluorescence intensity were visible. We conclude from these observations that oocysts labeled with 1 mg/ml NBD-C_6-HPC may be saturated with dye, and that dye saturation may mask differences in dye uptake between oocyst populations.

Effect of Chlorine Experiment 2. The experiment to assess the effect of different concentrations of NBD-C_6-HPC (Figure 3.54) on oocyst labeling indicated that 1 mg/ml may be excessive for quantitative studies, and may result in saturation and possible quenching of fluorescence. Based on this observation, a new chlorine treatment of genotype II oocysts was performed and the oocysts labeled with 10 µg/ml NBD-C_6-HPC, instead of the 1 mg/ml concentration used previously. Prior to labeling, the oocysts were incubated in 1% SDS, but were not boiled. FL1 fluorescence profiles showed the presence of a population of oocysts fluorescing about 40 times more brightly than untreated oocysts (Figure 3.55). The proportion of oocysts in this subpopulation increased with increasing chlorine dose (Figure 3.55, right column) and was absent in control oocysts. Plots of forward scatter vs FL1 showed that the brightly fluorescing oocysts, which appeared as a result of the chlorine exposure, also differed in their scatter properties (Figure 3.55, left column) indicating a reduction in size. These results, similar to those from a previous chlorine treatment (see Figure 3.51), indicate that within an oocyst population, some oocysts are more readily labeled. As a result of chlorination, a subpopulation which approaches 50% of the total after exposure to a dose of 4000 mg/min/l, becomes permeable to NBD-C_6-HPC. The fluorescence intensity of these oocysts is comparable to that of boiled oocysts (see Figure 3.51), suggesting that these oocysts have become permeable to NBD-C_6-HPC as a consequence of chlorination. Although no infectivity studies were performed as part of this project, a tentative explanation of these observations is that brightly fluorescing oocysts are no longer infectious due to the disruption of the oocyst wall.

Summary

In summary, to investigate the effect of two disinfectants on the oocyst wall, *C. parvum* genotype I and genotype II oocysts were exposed to different doses of ozone and chlorine. Doses were chosen such that various levels of inactivation were achieved. Inactivation of genotype I and genotype II oocysts exposed to the highest ozone dose of 140 mg/min/l was confirmed in cell culture. The effect of ozone and chlorine (highest dose = 4000 mg/min/l) was probed using three methods: 1) reactivity with antibodies specific for oocyst wall antigens detected by mAbs 2C3/Cellabs and COWP, as detected by polyclonal antibody anti-cpRL3, 2) binding of two fluorescent lectin-FITC conjugates to oocyst carbohydrates, and 3) incorporation of a fluorescent phospholipid probe. Binding of the three types of fluorescent probes was assessed qualitatively by immunoblotting or by microscopy and quantitated by flow cytometry. Chlorine treatment reduced binding of the 2C3/Cellabs mAbs but not that of anti-cpRL3. Ozone had no detectable effect on 2C3/Cellabs or anti-cpRL3 binding. Binding of *Maclura pomifera* agglutinin (MPA) to genotype I and genotype II oocysts was reduced by about 20% in response to ozone, whereas binding of *Helix pomatia* agglutinin (HPA) was decreased with genotype II oocysts but increased with type I. With respect to the uptake of labeled phospholipid, ozone and chlorine exposure resulted in the emergence of a brightly labeled oocyst subpopulation. The relative magnitude of this subpopulation was dose dependent and the fluorescence emitted by these oocysts approximated that emitted by heat inactivated oocysts. Together, these results suggest that ozone and chlorine do not affect the integrity of the oocyst carbohydrates, but that these disinfectants permeabilize the oocyst wall, facilitating the uptake of phospholipids into the oocysts and their incorporation into the oocysts.

CONCLUSIONS

In conclusion, in this project probes were used to identify and assess the relative roles of proteins, carbohydrates and lipids in the overall integrity and resistance of the *Cryptosporidium* oocyst wall for the assessment of disinfection of the oocyst. These included antibody probes anti-cpRL3 and mAbs 2C3/Cellabs for protein components COWP and 2C3/Cellabs antigen, lectins (particularly HPA and MPA) for carbohydrates and fluorogenic lipid probes (particularly NBD-C_6-HPC) for lipids. These probes were used to study the effect of disinfectants such as chlorine and ozone on oocyst wall components. The gene encoding the protein recognized by the mAb 2C3/Cellabs could not be cloned and studies to determine the effects of water treatment processes on components of the oocyst wall could not be performed. These studies are ongoing. Taken all together, knowledge gained from these studies will be of value to water utilities and regulatory agencies in order to improve current approaches to water purification and to design new strategies.

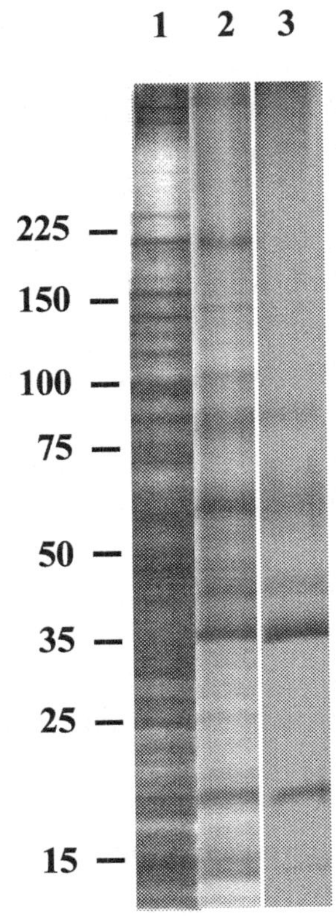

Figure 3.1 Analysis of oocyst wall proteins by SDS PAGE. 10^6/lane GCH1 whole oocysts (1), oocyst walls (2) and SDS-treated oocyst walls (3) were subjected to SDS PAGE under reducing conditions and stained with silver stain for total proteins.

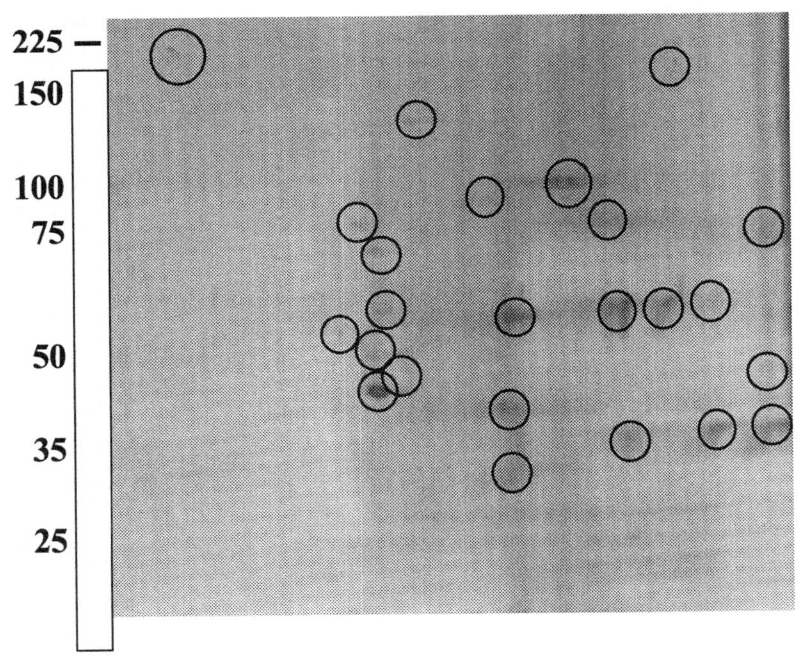

Figure 3.2 2D gel electrophoresis of GCH1 oocyst walls. Percoll-purified, SDS-treated oocyst walls were analyzed by 2D gel electrophoresis under reducing conditions and silver staining. Major spots are shown with circles.

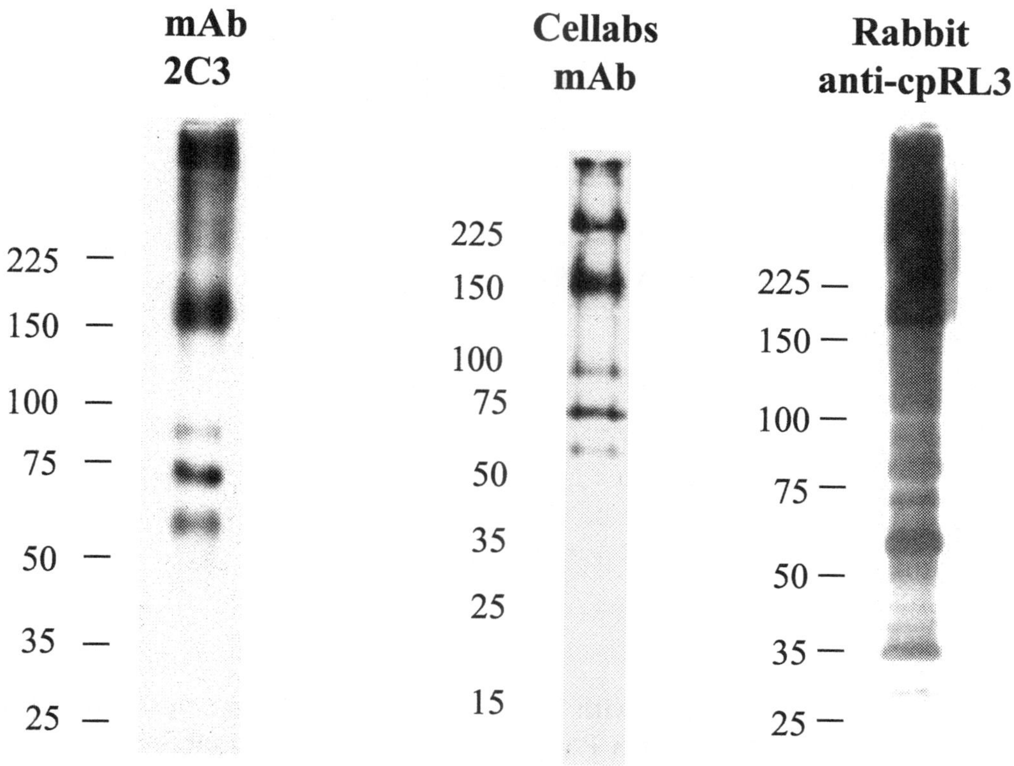

Figure 3.3 Reactivity of SDS-treated oocyst walls with mAb 2C3, Cellabs mAb and anti-cpRL3. 10^6/lane (2C3 or Cellabs) or 10^7/lane (anti-cpRL3) SDS-treated oocyst walls were analyzed by SDS PAGE and immunoblotting with the respective antibodies.

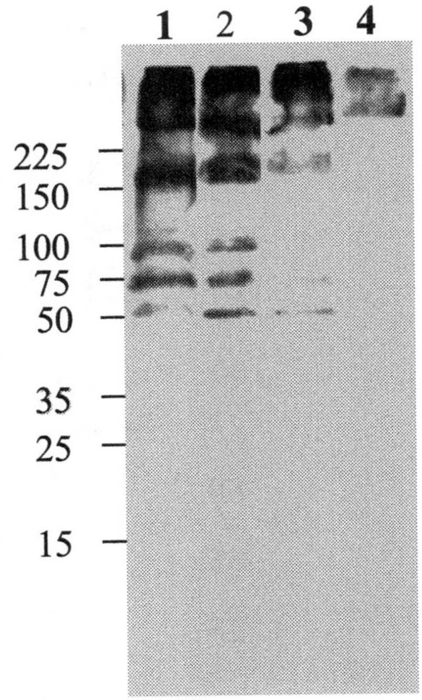

Figure 3.4 Reactivity of MAb 2C3 with intact oocysts, oocyst walls and SDS-treated oocyst walls. Proteins were resolved by SDS PAGE, transferred to nitrocellulose, and probed with mAb 2C3. Lane 1, intact oocysts; lane 2, oocyst walls; lane 3, proteins eluted from SDS-treated oocyst walls; lane 4, SDS-treated oocyst walls.

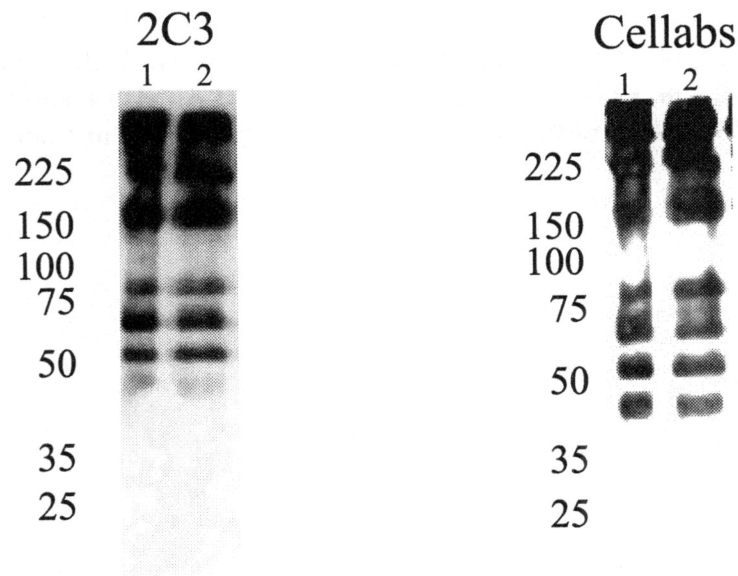

Figure 3.5 Effect of reducing agents on reactivity of mAbs 2C3 and Cellabs with oocyst walls. Oocyst walls were subjected to 5-15% gradient SDS PAGE under non-reducing (lane 1) or reducing (lane 2) conditions, transferred to nitrocellulose and probed with MAb 2C3 or Cellabs mAb.

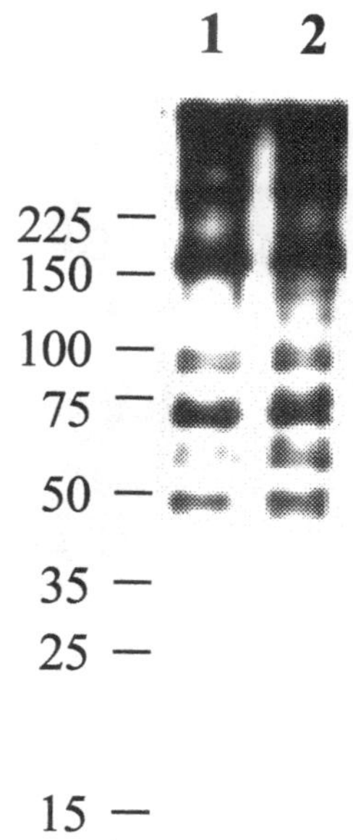

Figure 3.6 Effect of mild periodate treatment on reactivity of MAb 2C3 with oocyst walls. Oocyst walls were subjected to 5-15% gradient SDS PAGE and transferred to nitrocellulose. Strip 2 was subjected to mild periodate treatment and Strip 1 treated with buffer alone under the same conditions. Both strips were probed with mAb 2C3.

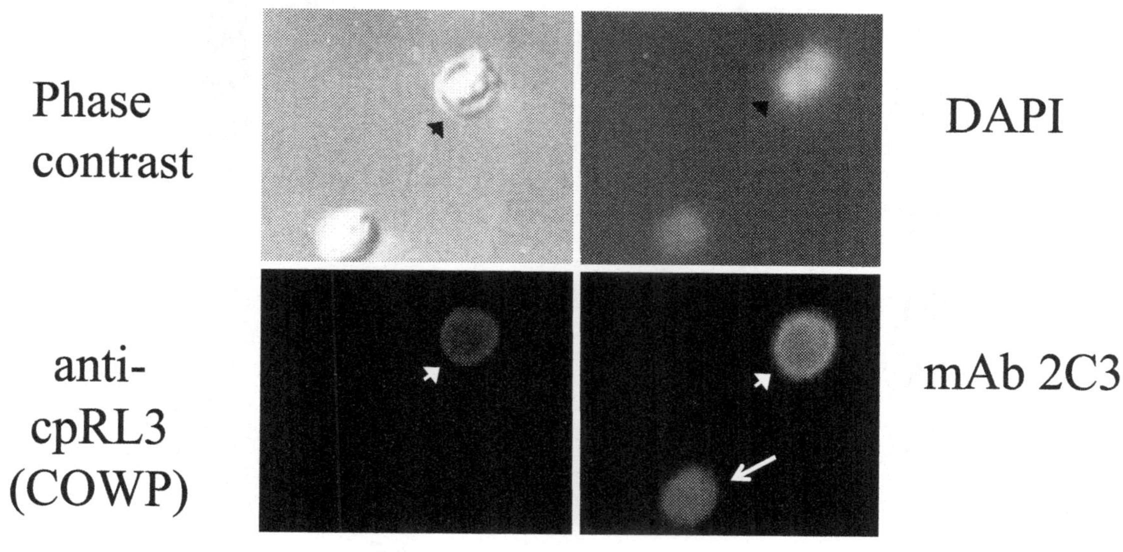

Figure 3.7 Reactivity of intact and disrupted oocysts with anti-cpRL3 and MAb 2C3. Arrowhead indicates disrupted oocyst. Anti-cpRL3 reacts with an inner oocyst wall component whereas mAb 2C3 recognizes epitopes exposed on the intact oocyst wall (arrow).

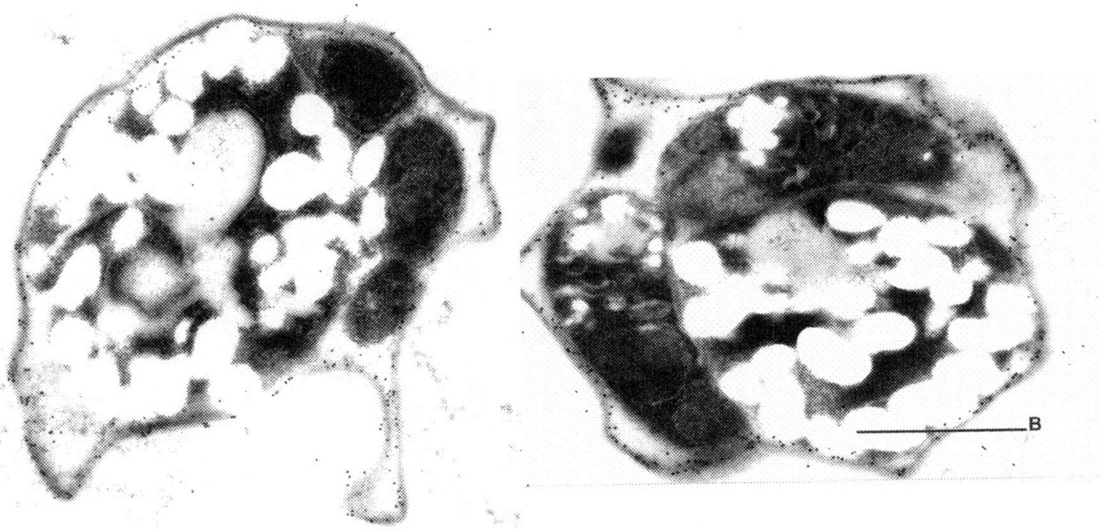

Figure 3.8 Transmission electron micrograph of purified oocysts labeled with mAb 2C3. MAb 2C3 binding to the outer oocyst wall is detected with gold labeled anti-mouse IgG. Size bar = 1 µm.

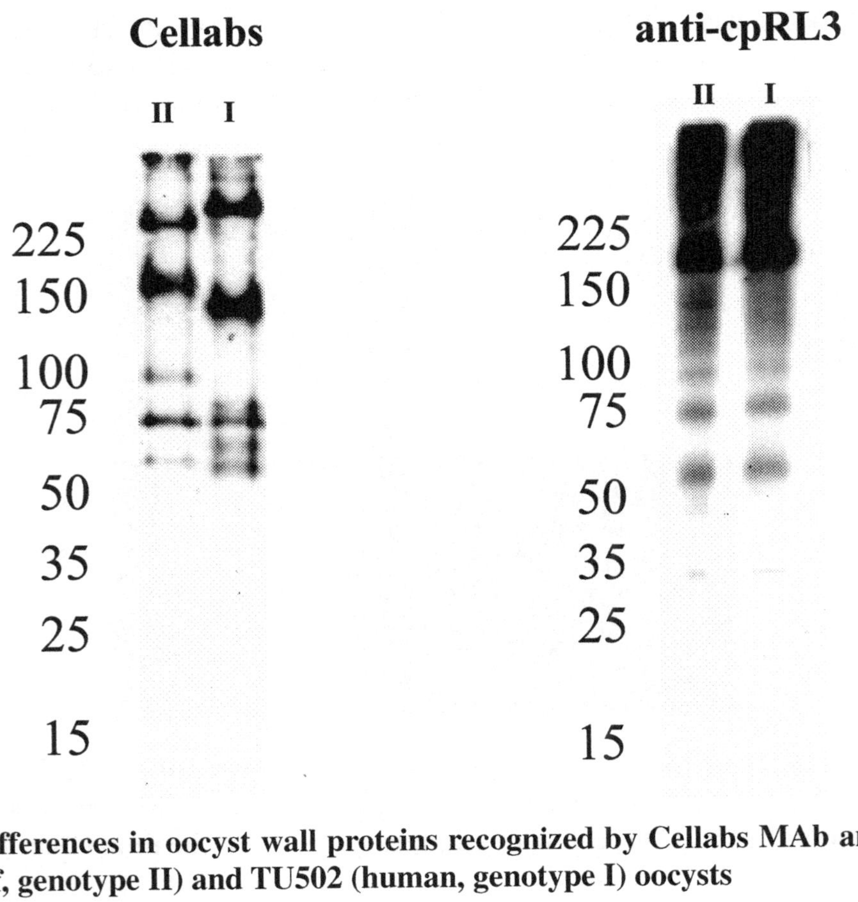

Figure 3.9 Differences in oocyst wall proteins recognized by Cellabs MAb and anti-cpRL3 in GCH1 (calf, genotype II) and TU502 (human, genotype I) oocysts

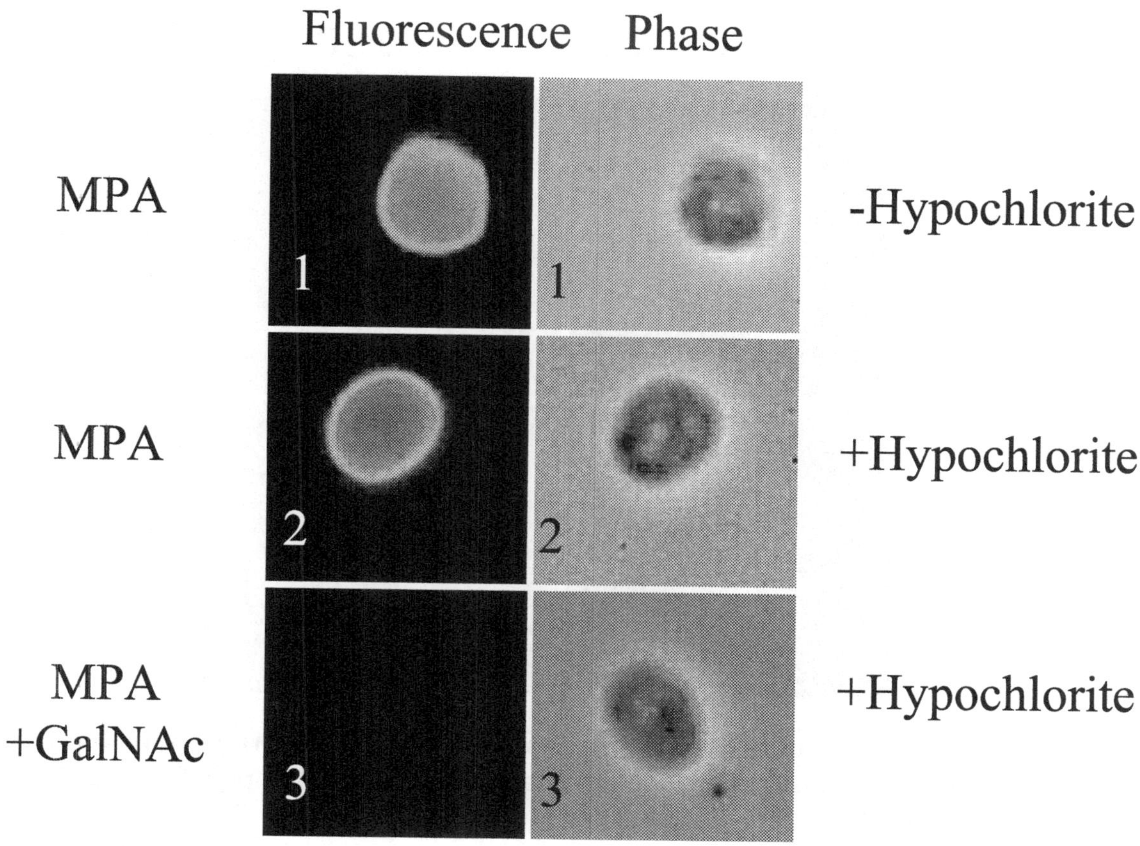

Figure 3.10 FITC-MPA binding to *C. parvum* oocysts. Intact oocysts were treated with PBS (1) or hypochlorite (2, 3), fixed onto poly-L-lysine-coated slides with 4% paraformaldehyde and stained with FITC-MPA (1, 2) or FITC-MPA, pre-incubated with GalNAc (3).

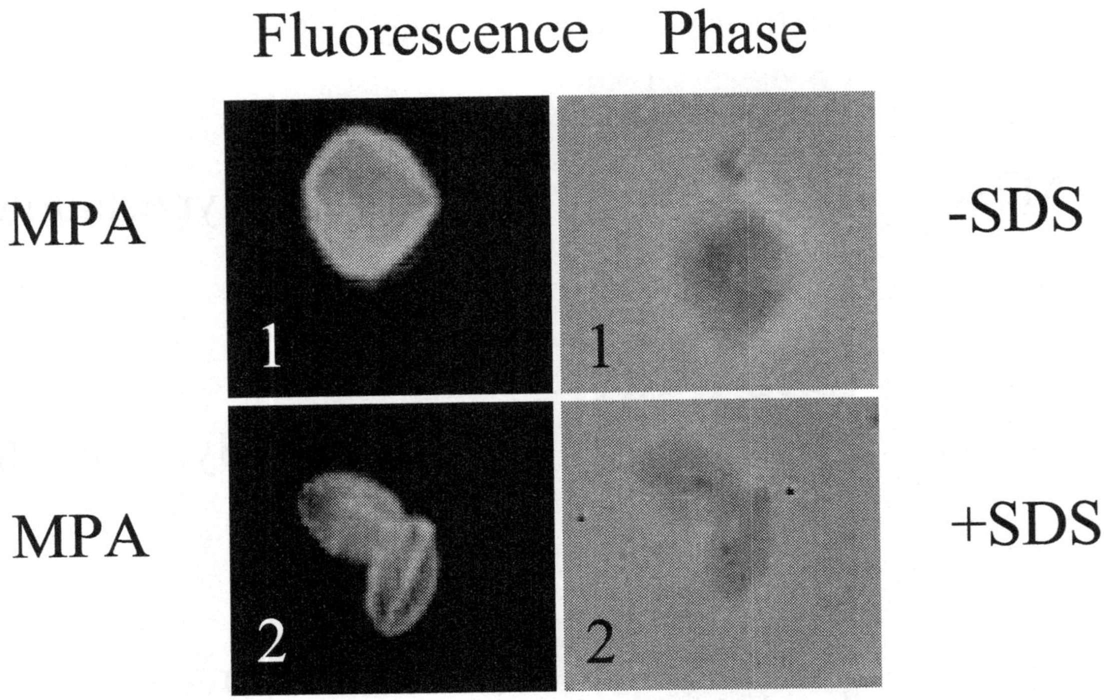

Figure 3.11 FITC-MPA binding to *C. parvum* oocyst walls. Oocyst walls were treated with PBS (1) or SDS (2), fixed onto poly-L-lysine-coated slides with 4% paraformaldehyde and stained with FITC-MPA (1, 2).

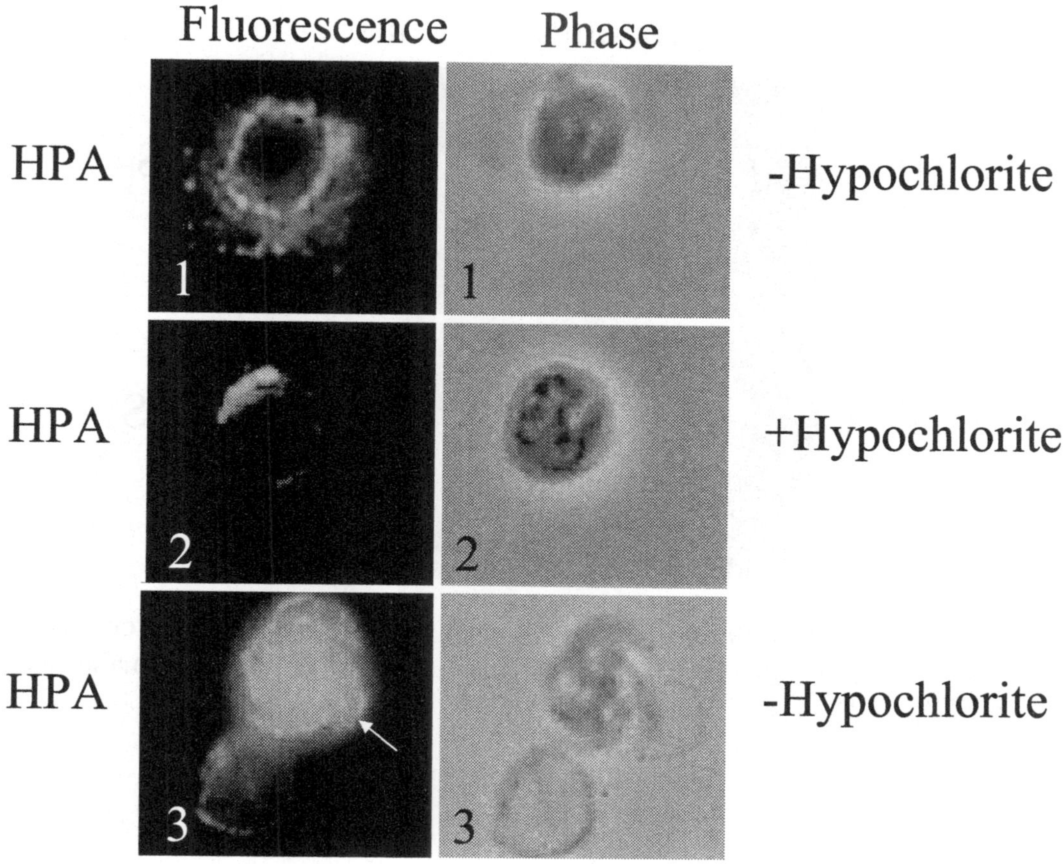

Figure 3.12 FITC-HPA binding to *C. parvum* oocysts. Intact (1, 2) or newly excysted (3) oocysts were treated with PBS (1, 3) or hypochlorite (2), fixed onto poly-L-lysine-coated slides with 4% paraformaldehyde and stained with FITC-HPA (1, 2, 3). Arrow indicates staining of sporozoites.

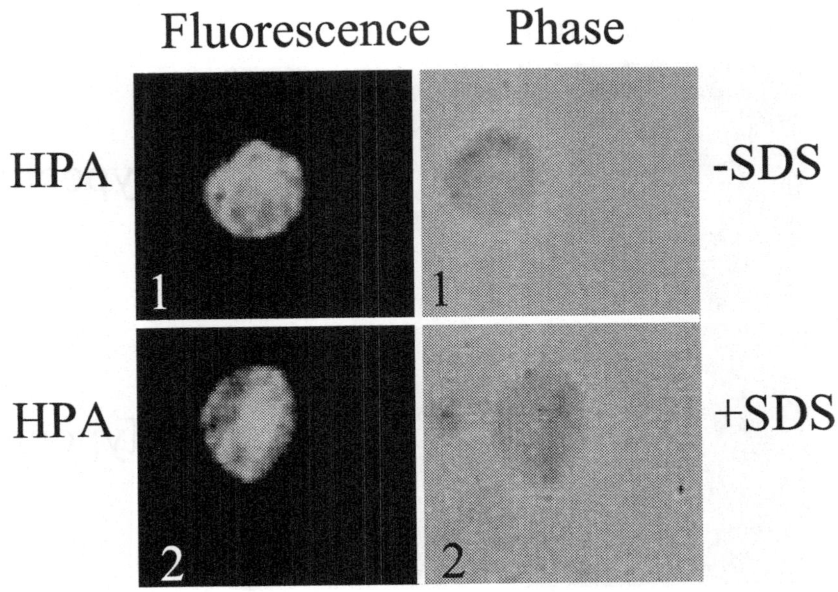

Figure 3.13 FITC-HPA binding to *C. parvum* oocyst walls. Oocyst walls were treated with PBS (1) or SDS (2), fixed onto poly-L-lysine-coated slides with 4% paraformaldehyde and stained with FITC-HPA (1, 2).

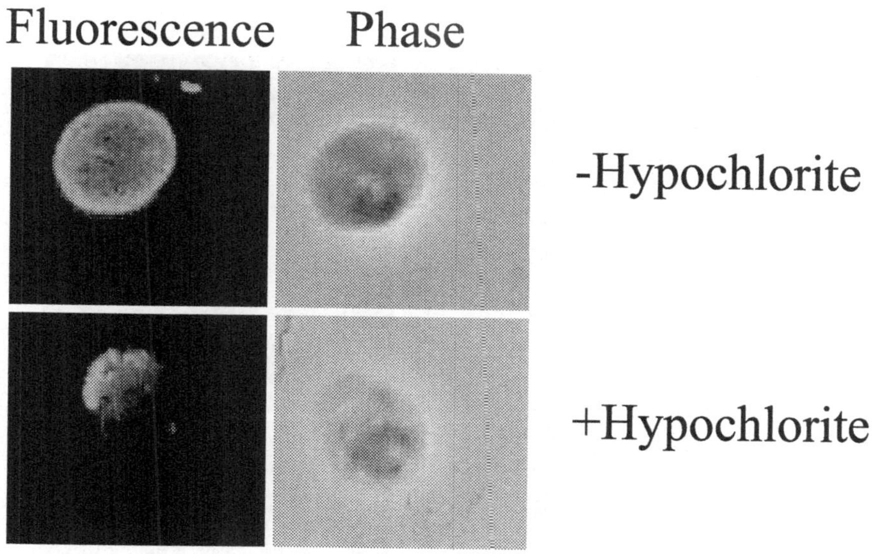

Figure 3.14 FITC-VVA binding to hypochlorite treated *C. parvum* oocysts. Intact GCH1 oocysts were treated with PBS or hypochlorite, fixed onto poly-L-lysine-coated slides with 4% paraformaldehyde and stained with FITC-VVA.

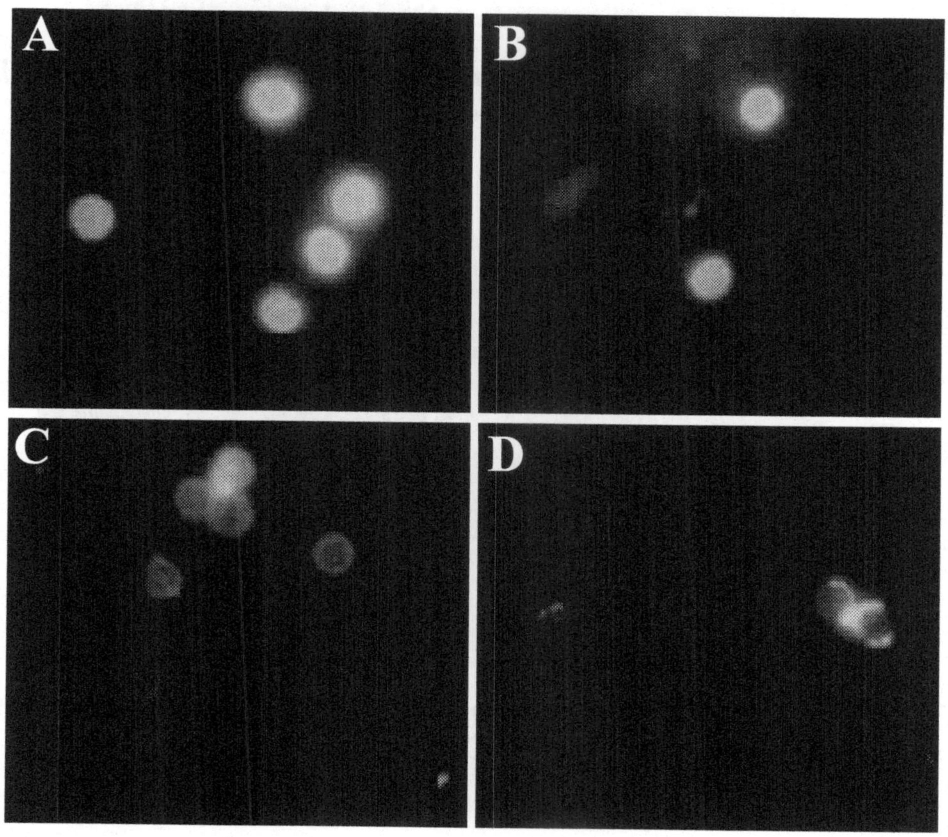

Figure 3.15 FITC-labeled lectin binding to fixed (4% paraformaldehyde) TU502 oocysts. A, FITC-MPA; B, FITC-AIA; C, FITC-PNA; D, FITC-Con A.

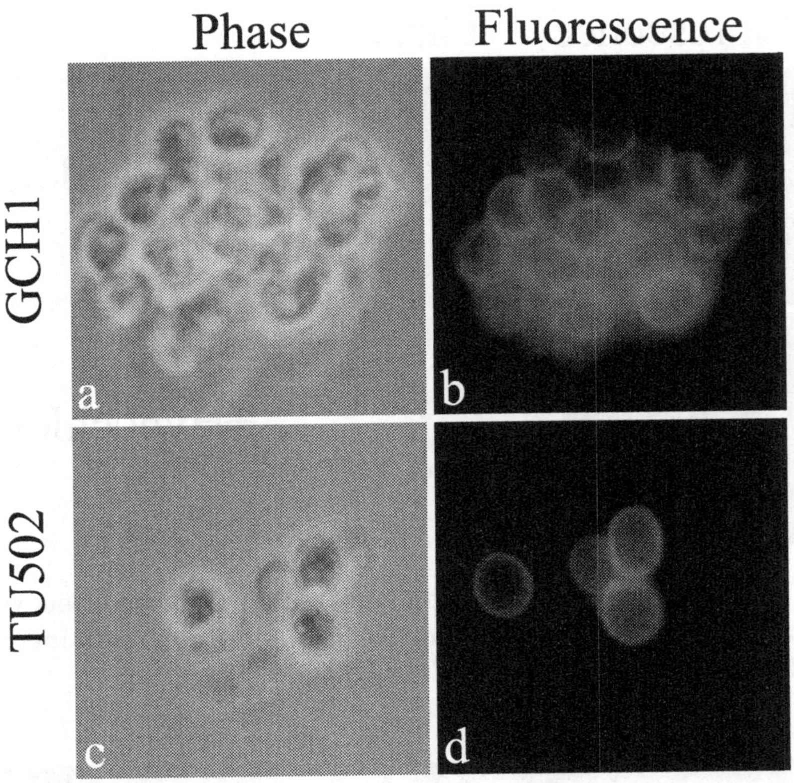

Figure 3.16 FITC-MPA binding to *C. parvum* oocysts in suspension. Intact GCH1 (a and b) or TU502 (c and d) oocysts in suspension were stained with FITC-MPA.

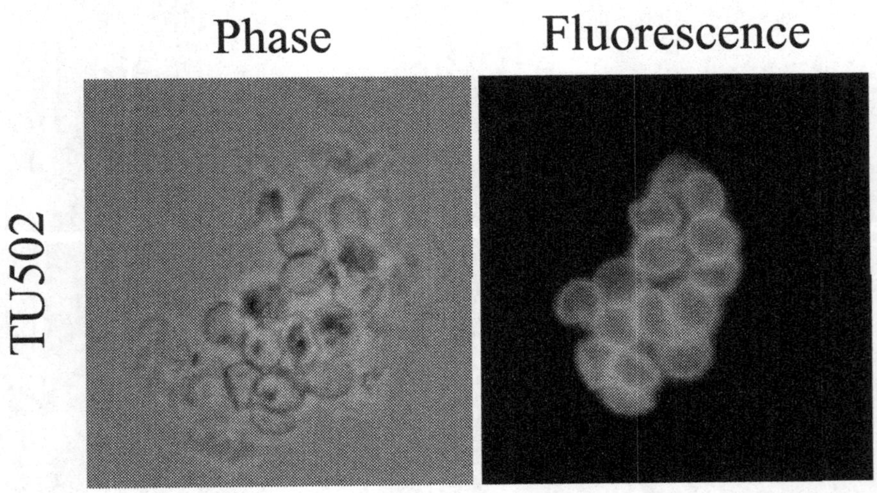

Figure 3.17 FITC-MPA binding to *C. parvum* TU502 oocysts in suspension. Intact TU502 oocysts in suspension were stained with FITC-MPA.

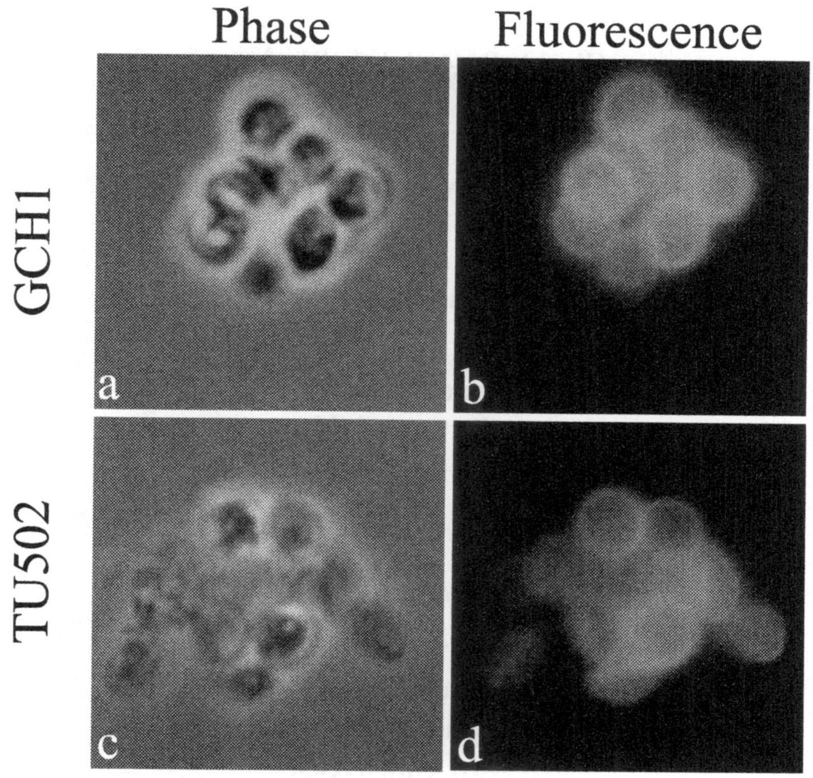

Figure 3.18 FITC-AIA binding to *C. parvum* oocysts in suspension. Intact GCH1 (a and b) or TU502 (c and d) oocysts in suspension were stained with FITC-AIA.

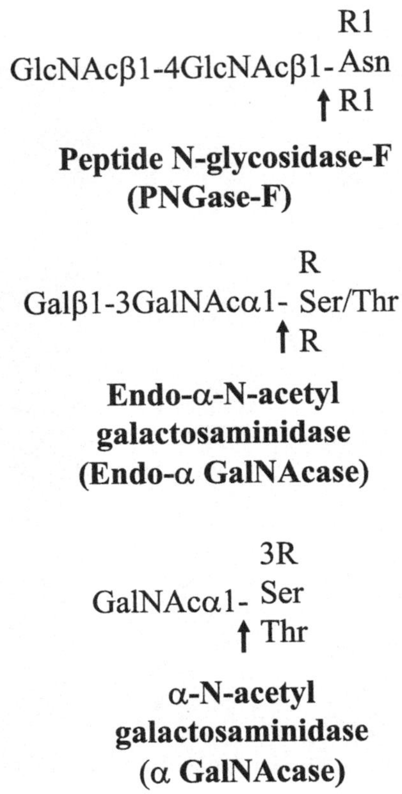

Figure 3.19 Glycosidases used for digestion of oocyst walls and their sites of action

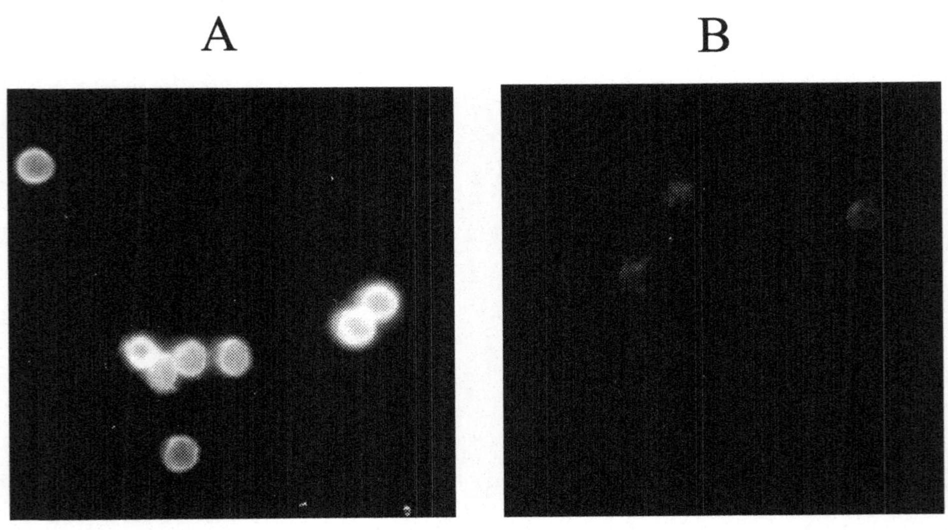

Figure 3.20 α-N-acetylgalactosaminidase treatment of oocysts reduces binding of FITC-MPA. Paraformaldehyde-fixed oocysts were treated with buffer alone (A) or with 5 mU/ml of a-N-acetylgalactosaminidase (B) for 18 hours at 37°C, washed and incubated with 1 µg/ml of FITC-MPA for 1 hr at room temperature and binding assessed by fluorescence microscopy.

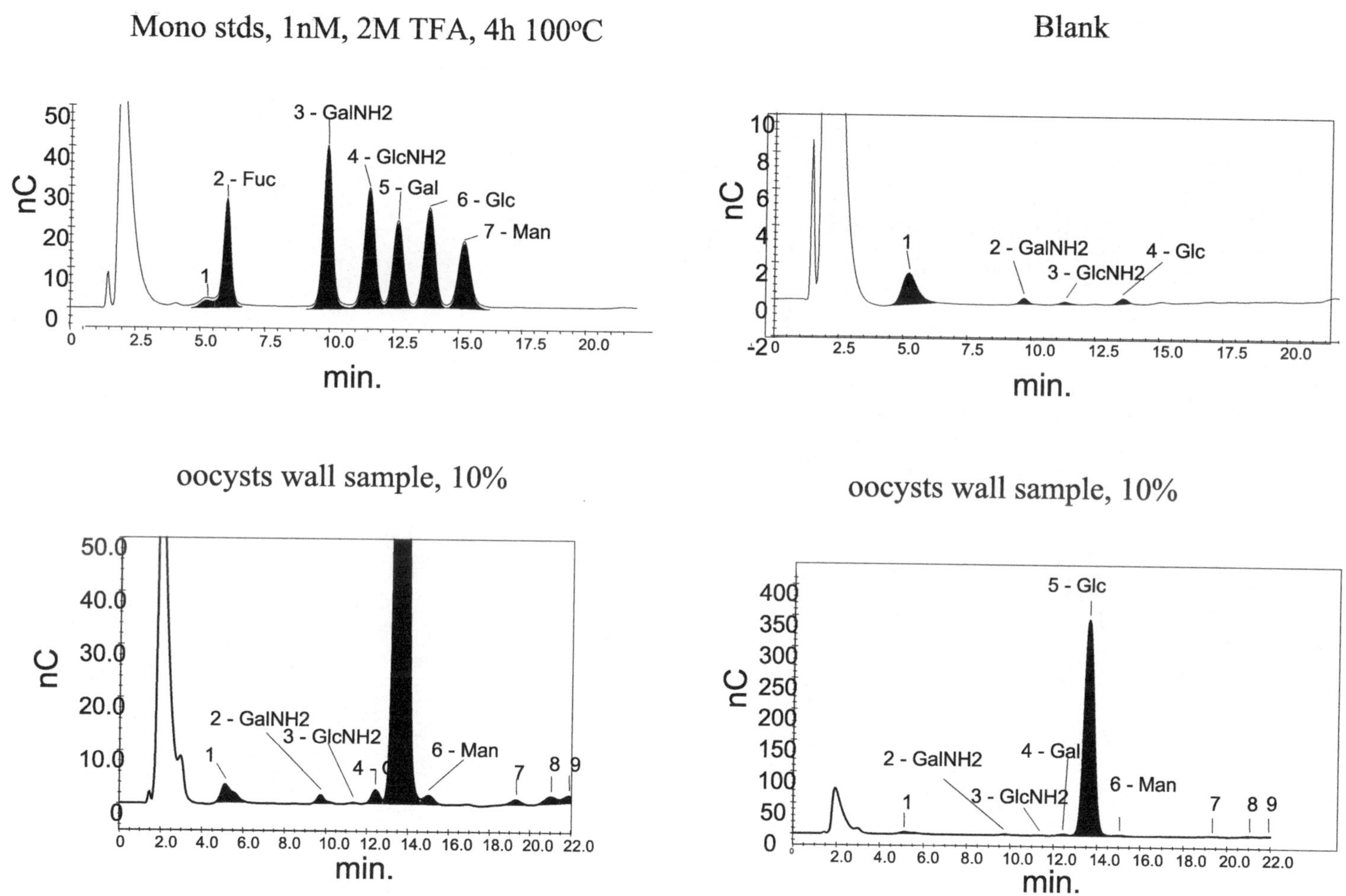

Figure 3.21 HPAEC analysis of *C. parvum* oocyst walls

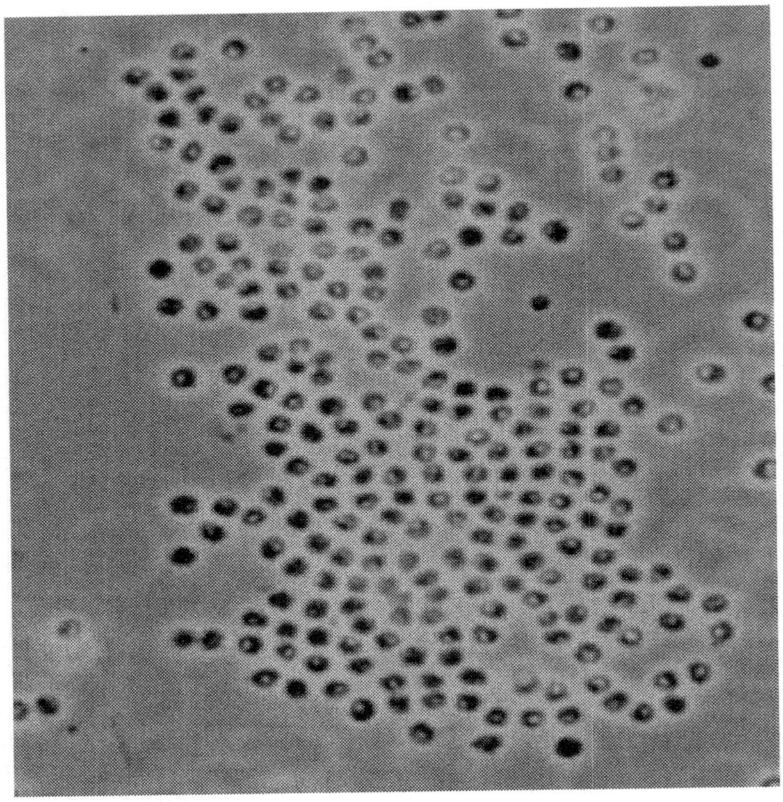

Figure 3.22 Phase-contrast micrograph of SDS-treated oocysts used for lipid probe studies. By phase microscopy the SDS-treated oocysts seemed to be intact but an inner membrane-like structure had retracted from the outer wall surface. This inner body refracted the light and it appeared to be disconnected completely from the oocyst wall.

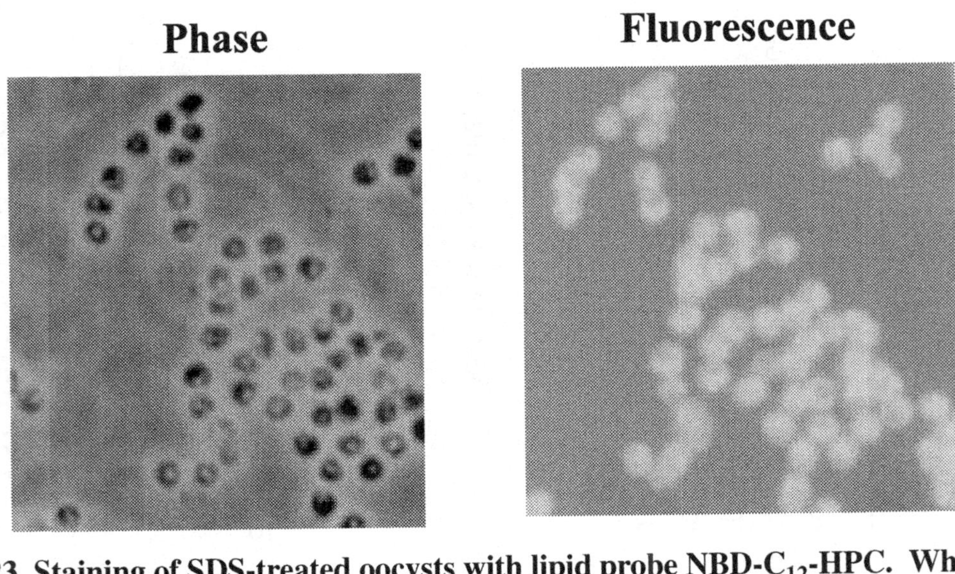

Figure 3.23 Staining of SDS-treated oocysts with lipid probe NBD-C_{12}-HPC. Whole oocysts were probed with NBD-C_{12}-HPC for 30 minutes at room temperature, followed by washing and photomicrography by phase contrast and fluorescence.

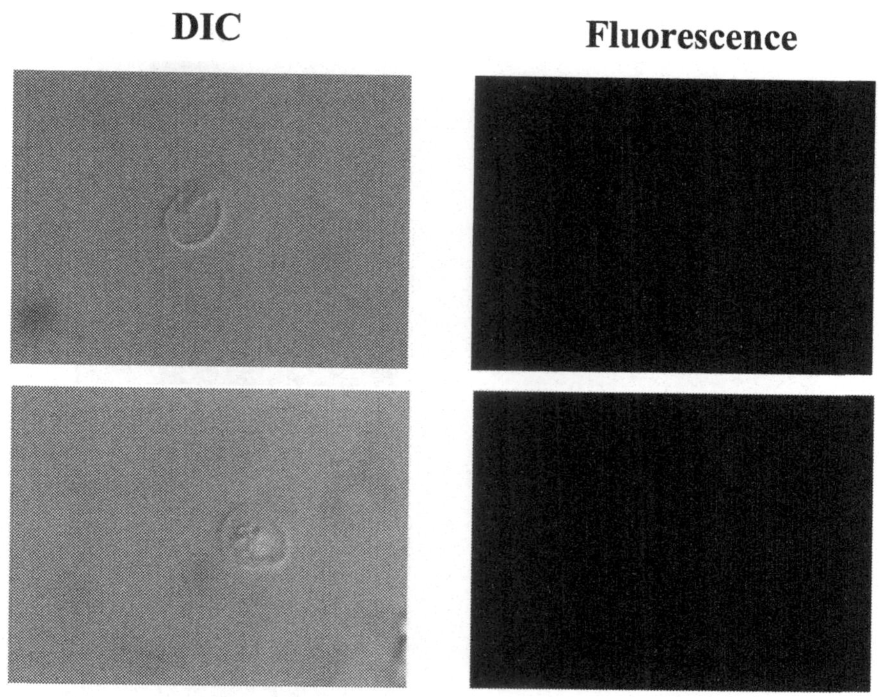

Figure 3.24 Non-treated, isolated oocyst walls do NOT incorporate fluorescent lipid probes. Isolated oocyst walls were probed with NBD-C_6-HPC for 30 minutes at room temperature, followed by washing and photomicrography by DIC and fluorescence.

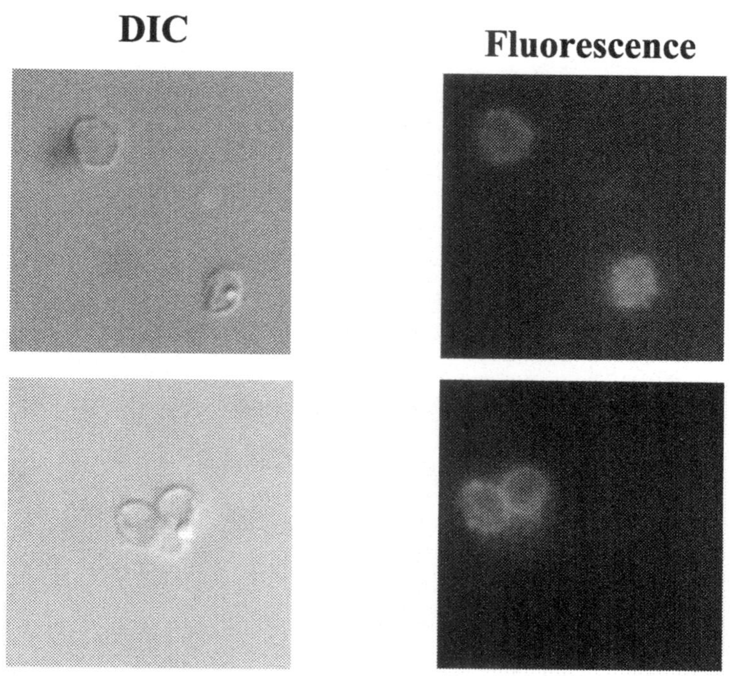

Figure 3.25 Isolated oocyst walls treated with SDS and heat incorporate fluorescent phospholipid probes. Isolated oocyst walls were probed with NBD-C_6-HPC for 30 minutes at room temperature, followed by washing and photomicrography by DIC and fluorescence.

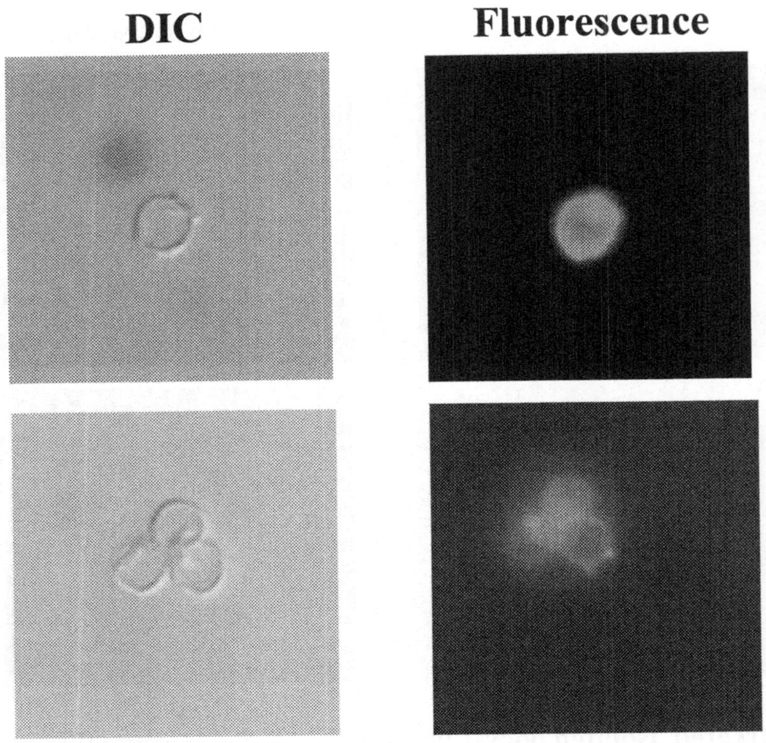

Figure 3.26 Isolated oocyst walls treated with SDS and heat incorporate fluorescent sphingolipid probes. Isolated oocyst walls were probed with NBD-C_6-sphingomyelin for 30 minutes at room temperature, followed by washing and photomicrography by DIC and fluorescence.

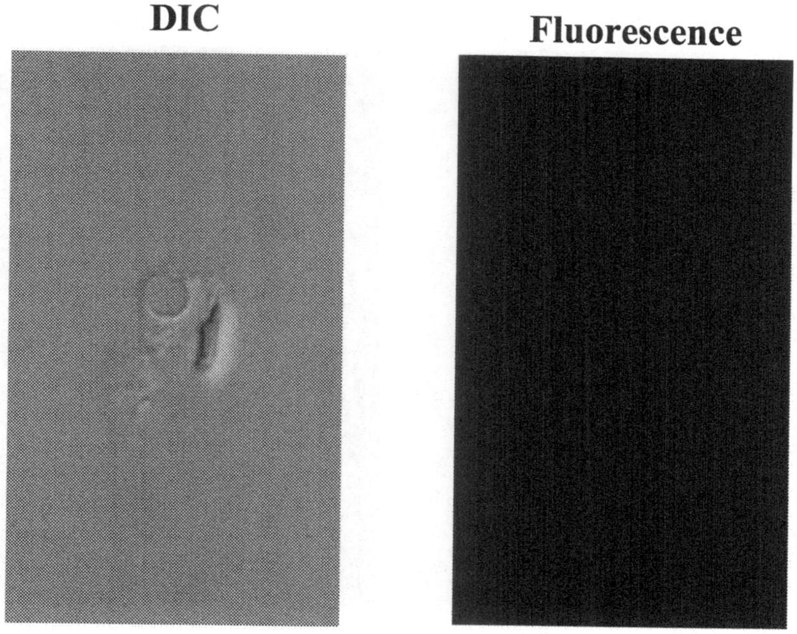

Figure 3.27 Isolated oocyst walls treated with SDS and heat do NOT incorporate fluorescent cholesterol probes. Isolated oocyst walls were probed with NBD-Sterol for 30 minutes at room temperature, followed by washing and photomicrography by DIC and fluorescence.

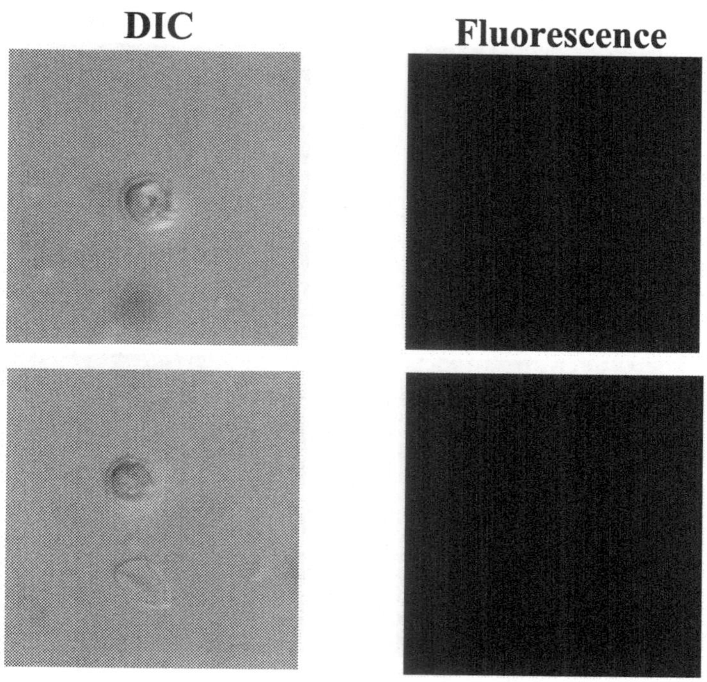

Figure 3.28 Non-treated, whole oocysts do NOT incorporate fluorescent lipid probes. Whole oocysts were probed with NBD-C_6-HPC for 30 minutes at room temperature, followed by washing and photomicrography by DIC and fluorescence.

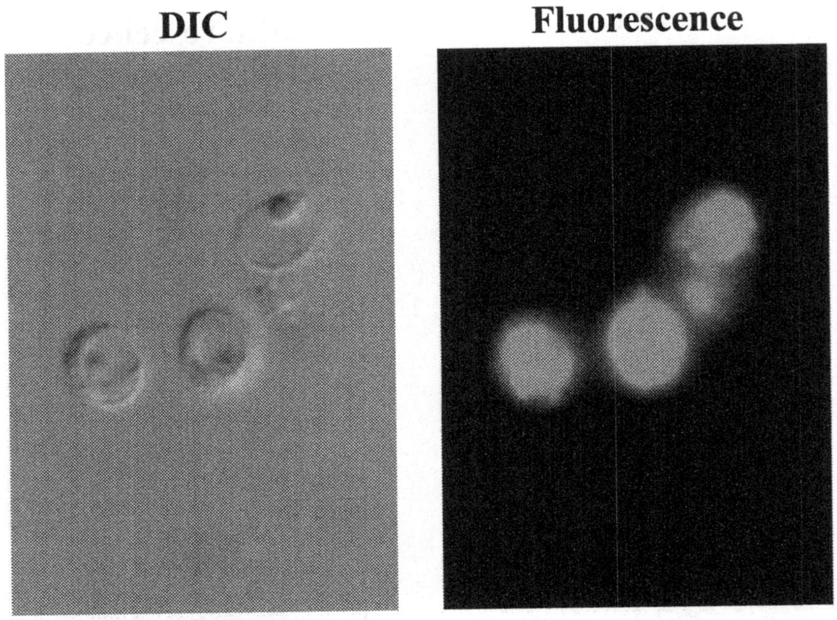

Figure 3.29 Whole oocysts treated with SDS and heat incorporate fluorescent phospholipid probes in both "internal structures" (i.e., collapsed sporozoite membranes) and oocyst walls. Whole oocysts were probed with NBD-C_6-HPC for 30 minutes at room temperature, followed by washing and photomicrography by DIC and fluorescence.

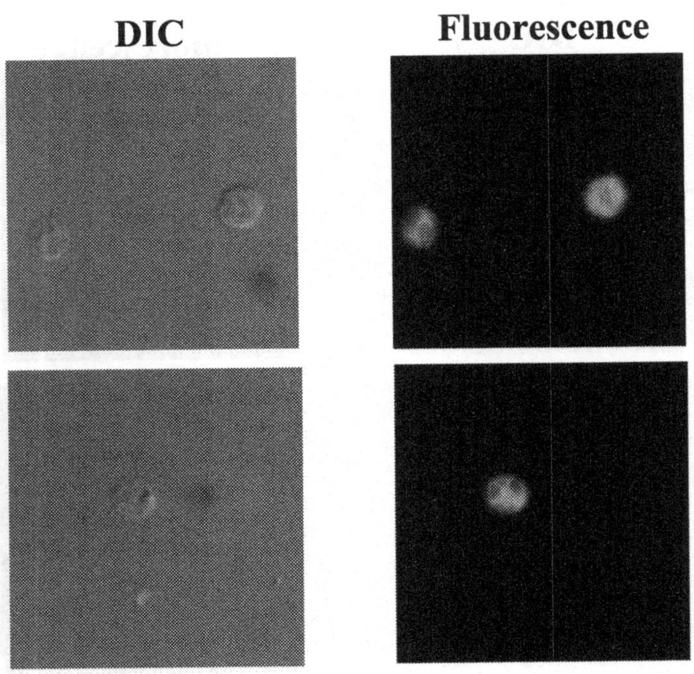

Figure 3.30 Whole oocysts treated with SDS and heat incorporate fluorescent cholesterol probes in "internal structures" (i.e., collapsed sporozoite membranes) but not in oocyst walls. Whole oocysts were probed with NBD-Sterol for 30 minutes at room temperature, followed by washing and photomicrography by DIC and fluorescence.

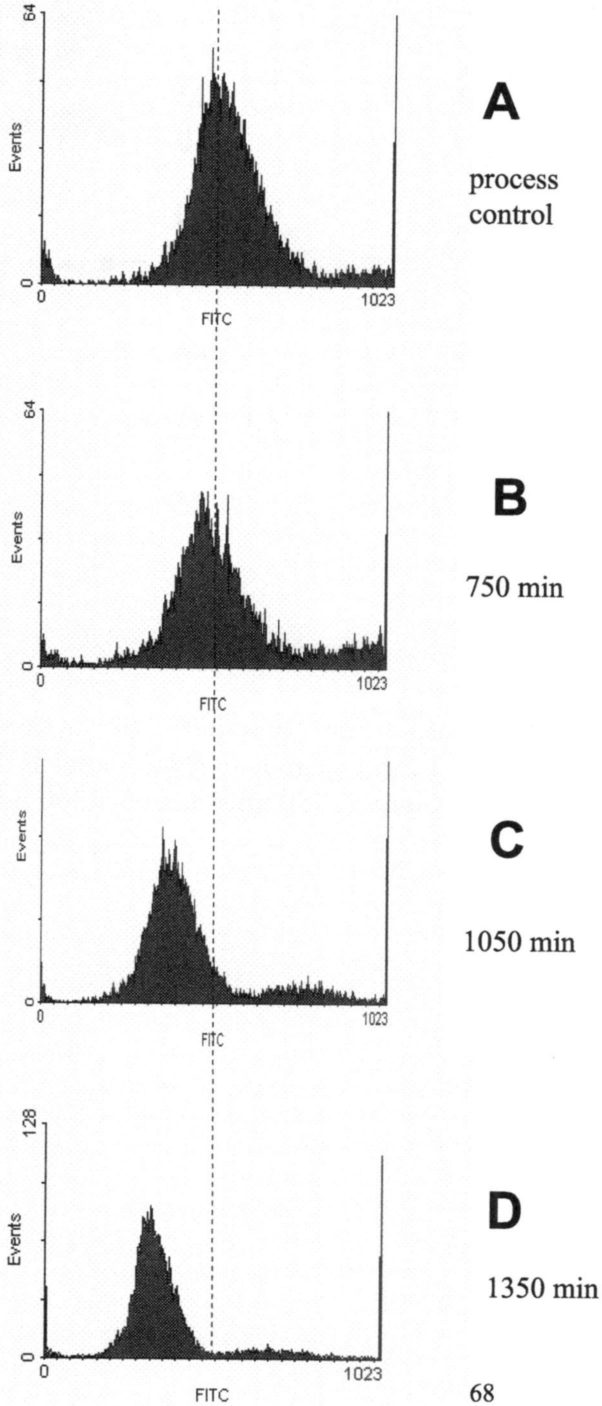

Figure 3.31 Flow cytometry analysis of 2C3 labeled oocysts of isolate GCH1 (genotype II) following chlorination. Chlorination conditions: 22°C; pH 7; initial chlorine residual, 4.25 mg/l.

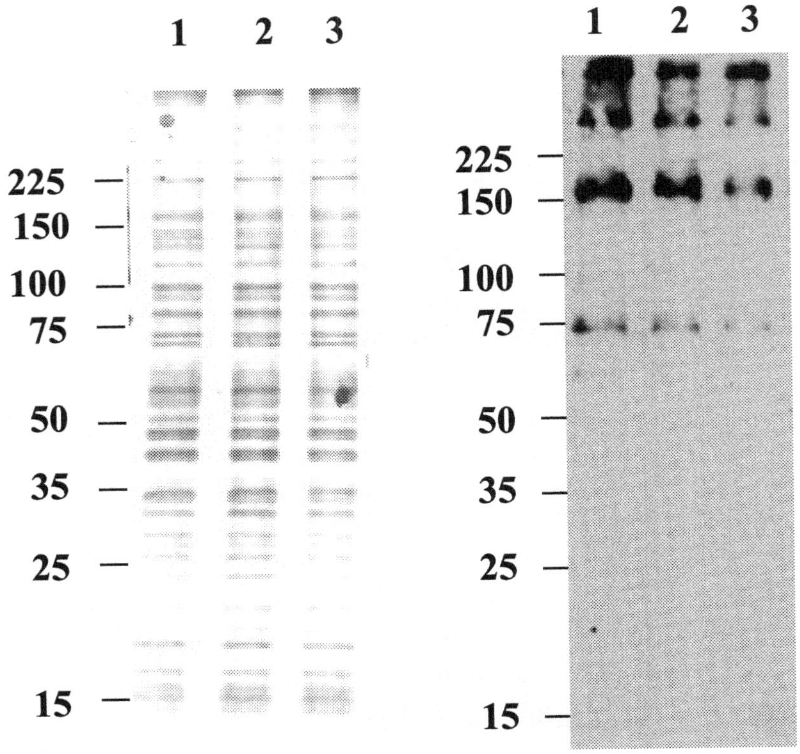

Figure 3.32 Effect of chlorination on reactivity of mAb 2C3 with oocysts by immunoblotting. Left panel shows total proteins; right panel is immunoblot probed with mAb 2C3. Lane 1, process control; lane 2, 750 min; lane 3, 1500 min.

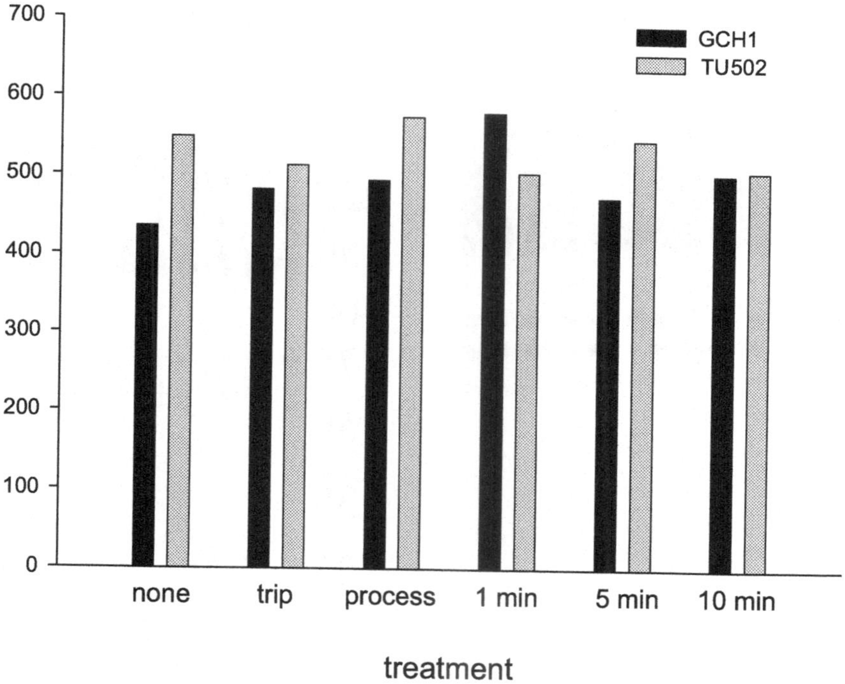

Figure 3.33 2C3 fluorescence of ozone treated *C. parvum* oocysts. Oocysts of isolates GCH1 (genotype II) and TU502 (genotype I) were exposed to 1.4 mg/l initial ozone concentration at room temperature and pH 7. 2C3 antibody reactivity was quantitated by flow cytometry. Y-axis shows peak fluorescence channel. None, no treatment; trip, trip control; process, process control; 1 min, 5 min, 10 min indicates time of exposure.

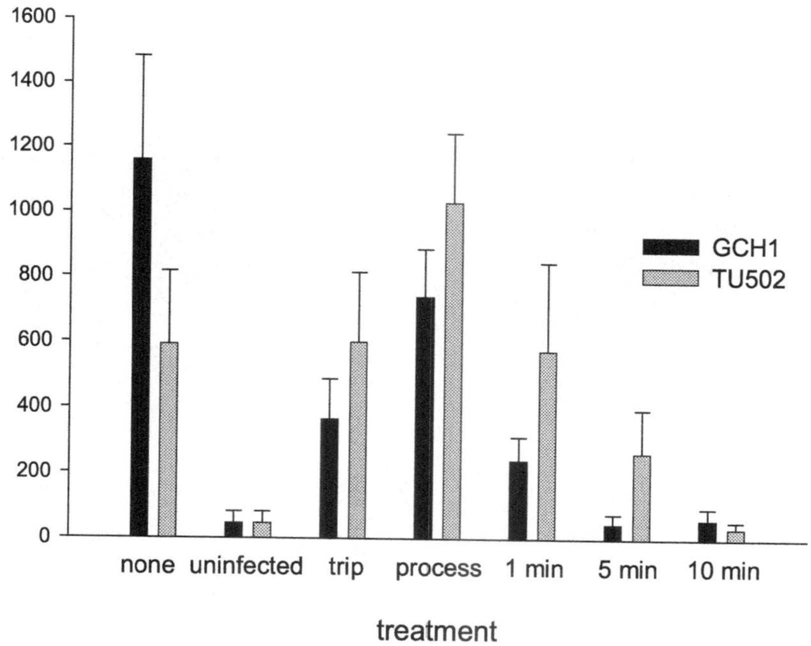

Figure 3.34 Cell culture infectivity of ozone treated *C. parvum* oocysts. Oocysts used for 2C3 reactivity experiments were tested for infectivity in MDBK cultures. The graph demonstrates a dose-dependent inactivation of oocysts.

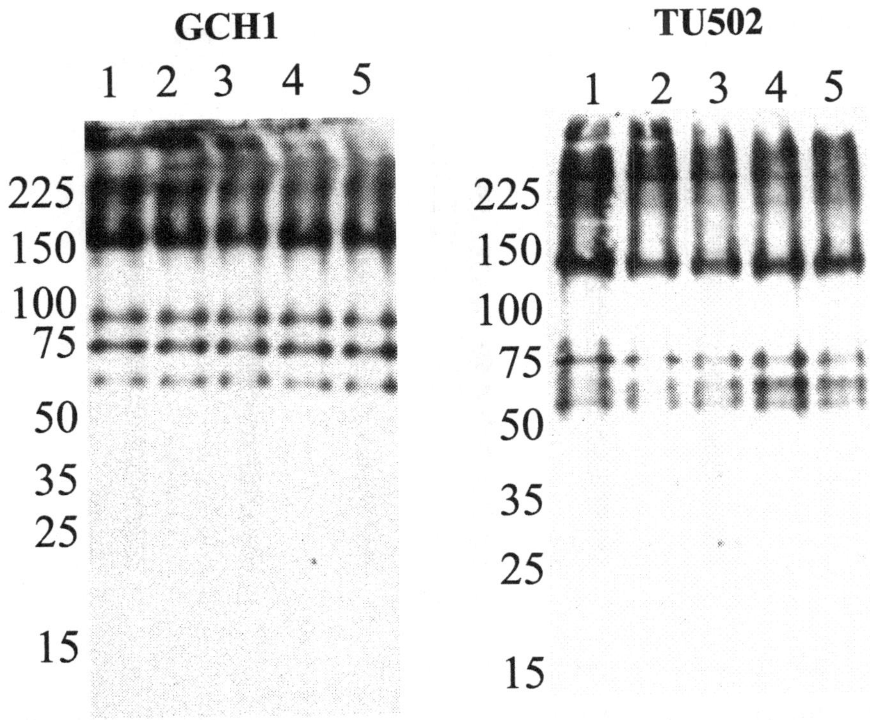

Figure 3.35 Effect of ozone treatment on reactivity of Cellabs mAb with GCH1 (calf, genotype II) and TU502 (human, genotype I) oocysts. Lane 1, trip control; lane 2, process control; lane 3, 1 min; lane 4, 5 min; lane 5, 10 min.

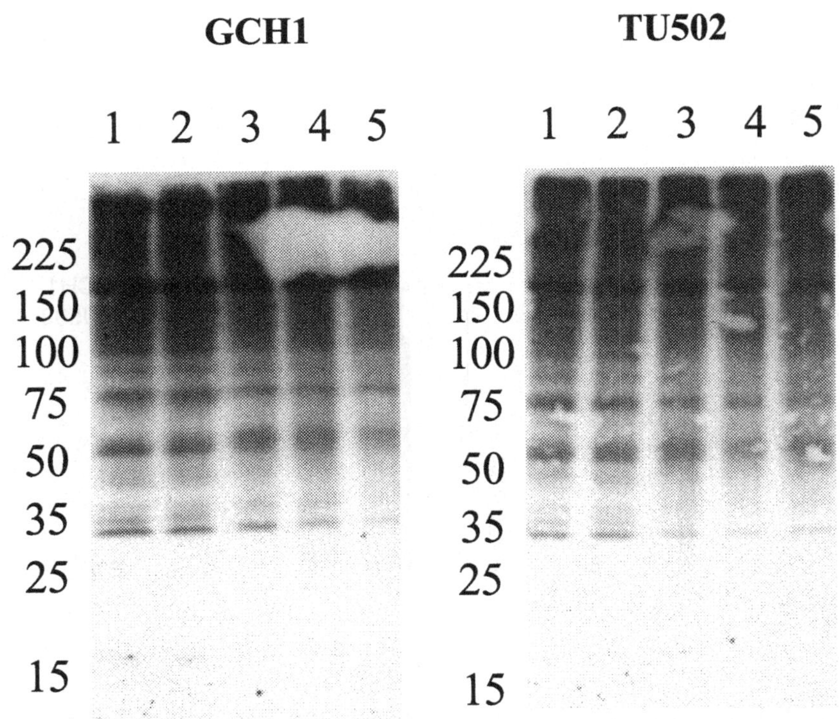

Figure 3.36 Effect of ozone treatment on reactivity of anti-cpRL3 with GCH1 (calf, genotype II) and TU502 (human, genotype I) oocysts. Lane 1, trip control; lane 2, process control; lane 3, 1 min; lane 4, 5 min; lane 5, 10 min.

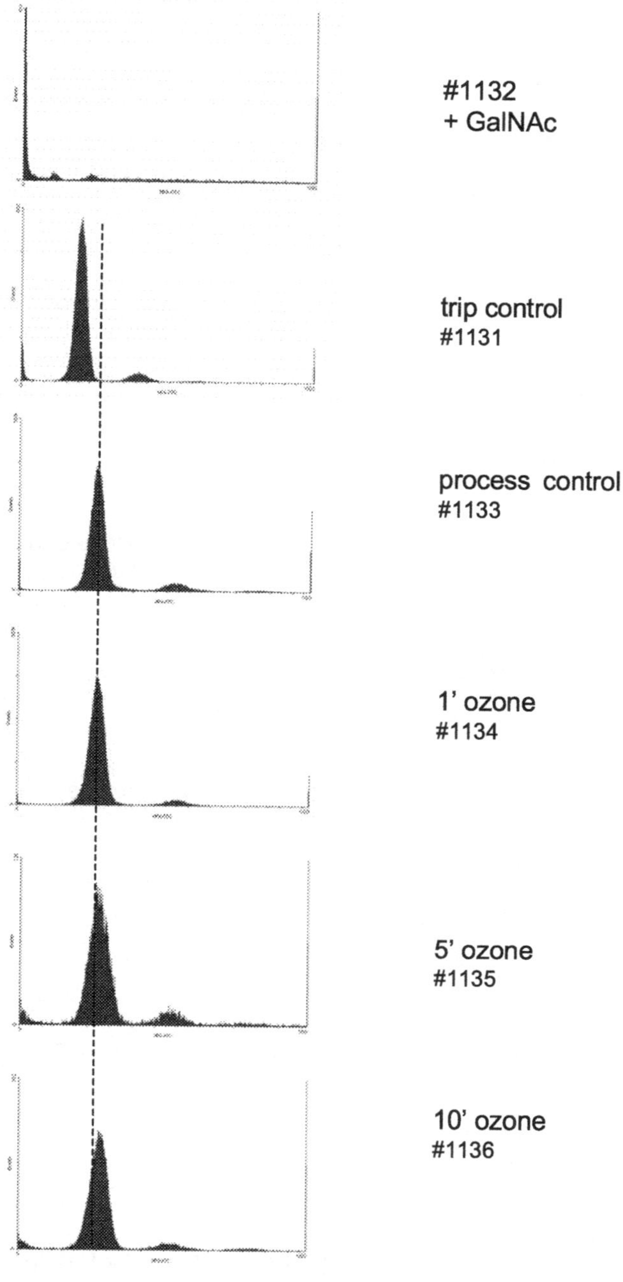

Figure 3.37 Flow cytometry analysis of FITC-MPA labeled oocysts of isolate GCH1 (genotype II). Paraformaldehyde fixed and methanol permeabilized oocysts were reacted overnight with lectin and analyzed. Shown are fluorescence (FL1 channel) histograms of 50,000 events. X-axis, fluorescence; Y-axis, number of events. File numbers (#) are for internal reference. Secondary peaks probably originate from aggregating oocyst. A GalNAc inhibited control labeling is shown uppermost. Ozonation conditions were 1.4 mg/l, pH 7 at room temperature.

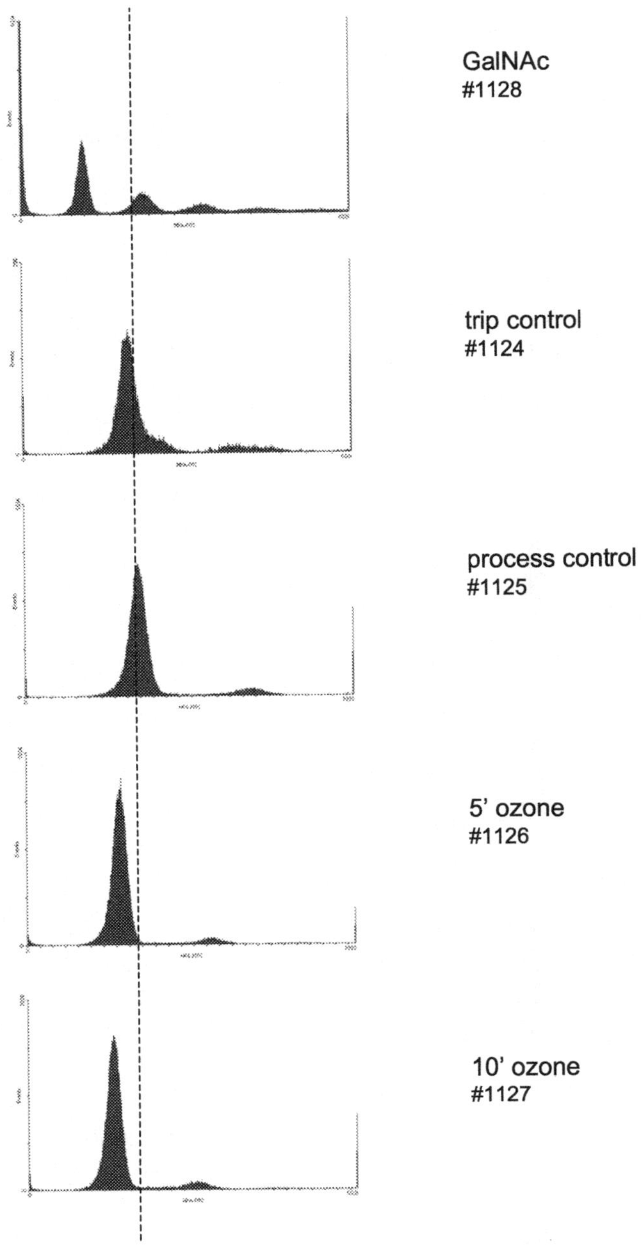

Figure 3.38 Flow cytometric analysis of FITC-MPA labeled oocysts of isolate TU502 (genotype I). Paraformaldehyde fixed and methanol permeabilized oocysts were reacted overnight with the FITC-lectin conjugate. Note a decrease in fluorescence in response to ozone. The magnitude of the ozone effect exceeds the difference between trip and process controls and is thus likely to reflect modification of carbohydrates resulting from the ozone. Ozonation conditions were 1.4 mg/l, pH 7 at room temperature.

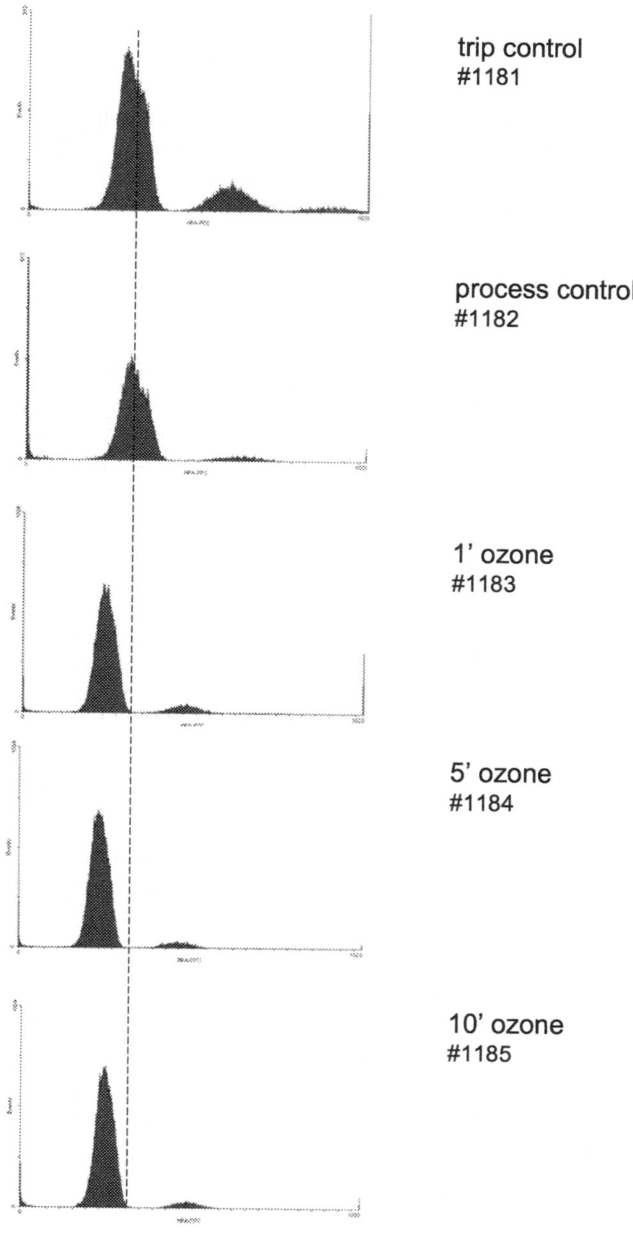

Figure 3.39 Flow cytometric analysis of FITC-MPA labeled GCH1 oocysts (genotype II). The experiment shown in Figures 3.37 and 3.38 was repeated with unfixed oocysts. Ozonation conditions were 1.4 mg/l, pH 7 at room temperature. Note reduction in fluorescence in ozone treated oocysts.

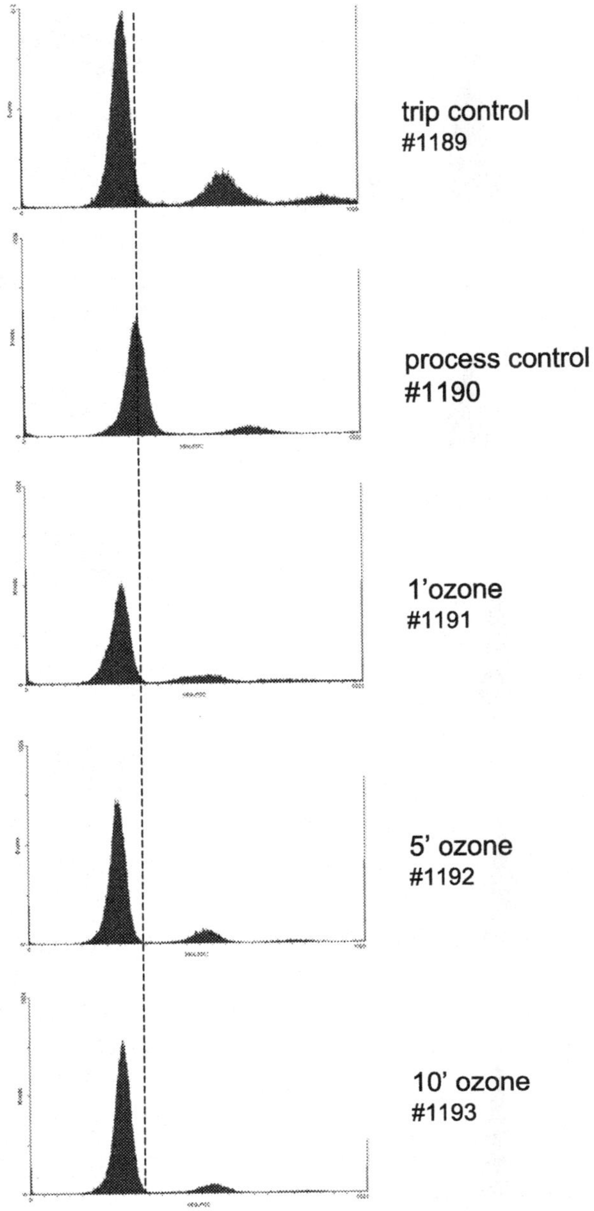

Figure 3.40 Flow cytometric analysis of FITC-MPA labeled TU502 oocysts (genotype I). See legend to Figure 3.39 for details. Treated oocysts were labeled without prior fixation. Ozonation conditions were 1.4 mg/l, pH 7 at room temperature.

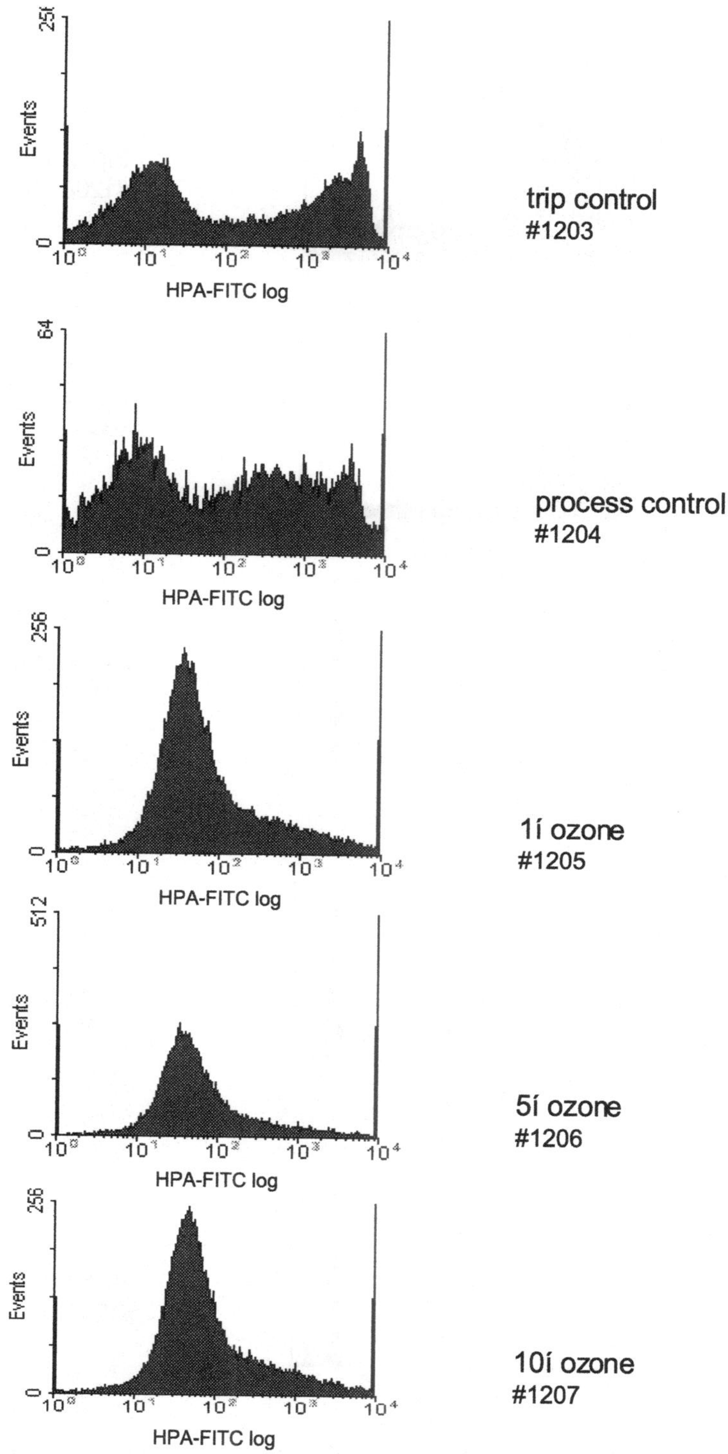

Figure 3.41 Flow cytometric analysis of ozone treated GCH1 oocysts (genotype II) reacted with FITC-HPA. Oocysts were fixed in paraformaldehyde and permeabilized with methanol prior to labeling. Trip and process controls showed two populations of oocysts, one intensity labeled, the other unlabelled. Ozone exposure eliminated reactivity to HPA. This effect does not appear to be dose dependent. Ozonation conditions were 1.4 mg/l, pH 7 at room temperature.

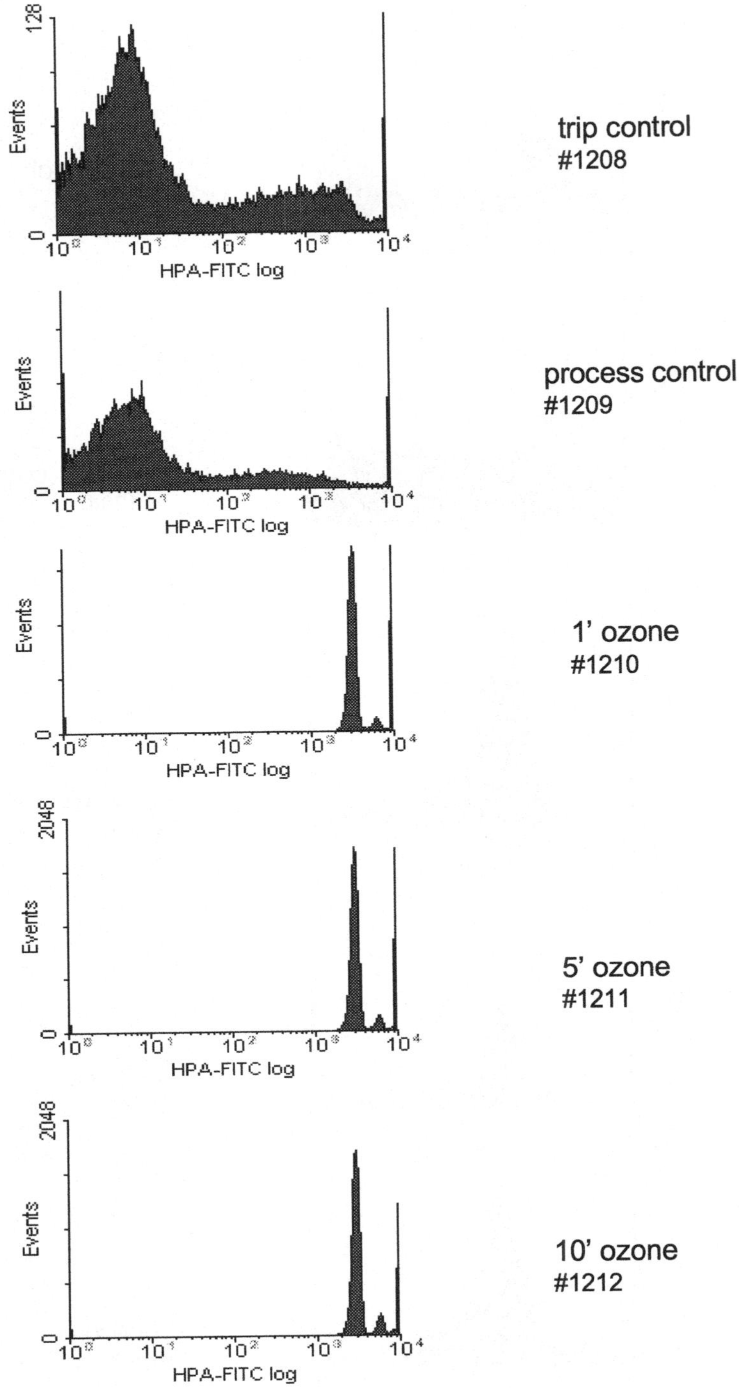

Figure 3.42 Flow cytometric analysis of ozone treated TU502 oocysts (genotype I) reacted with FITC-HPA. Oocysts were fixed and permeabilized as described in Figure 3.41. Note a large increase in lectin binding in the treated oocysts. Ozonation conditions were 1.4 mg/l, pH 7 at room temperature.

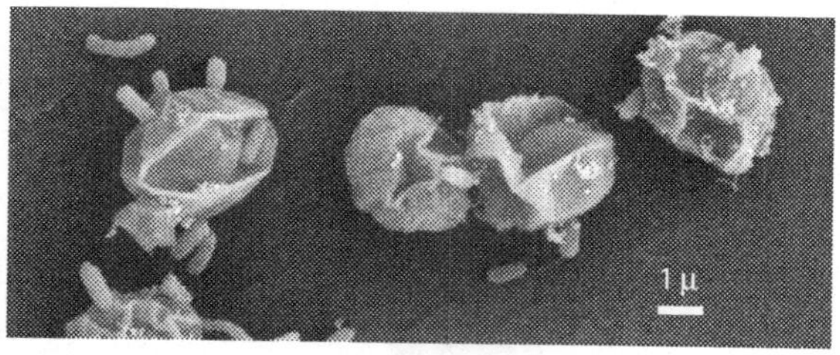

Process Control Without $Na_2S_2O_3$

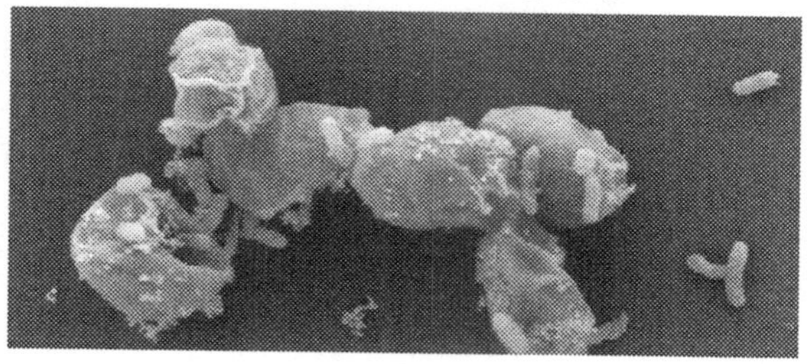

Process Control With $Na_2S_2O_3$

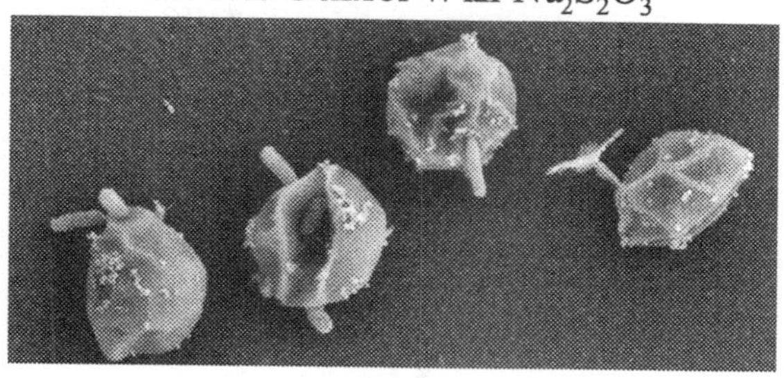

300 min

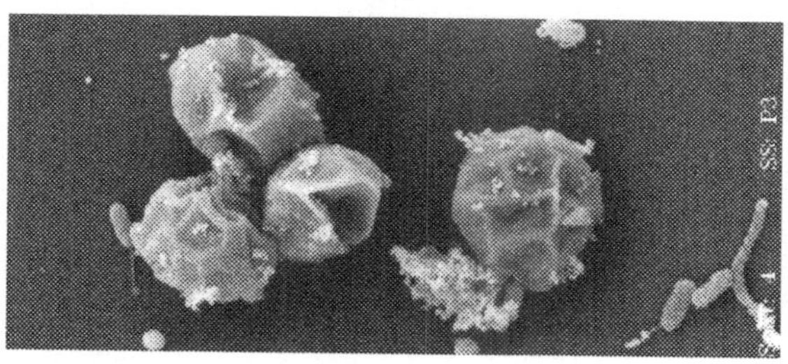

600 min

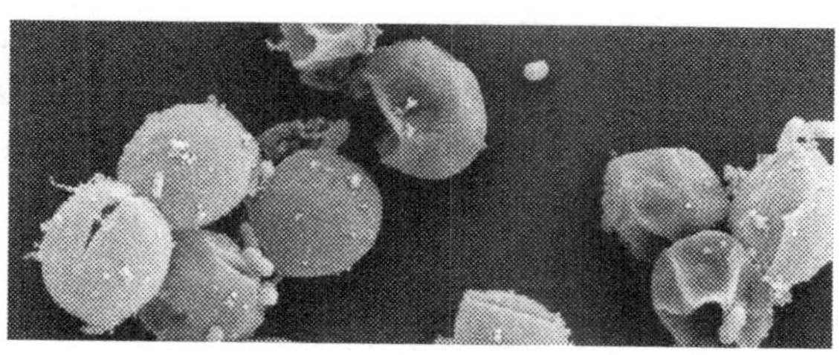

1200 min

Figure 3.43 Scanning electron micrographs of *C. parvum* oocysts exposed to 4 mg/ml chlorine. Oocysts from two process controls and three treatments (300 min, 600 min, and 1200 min) are shown. One micron size bar is shown in the top panel. The same amplification was applied to all micrographs.

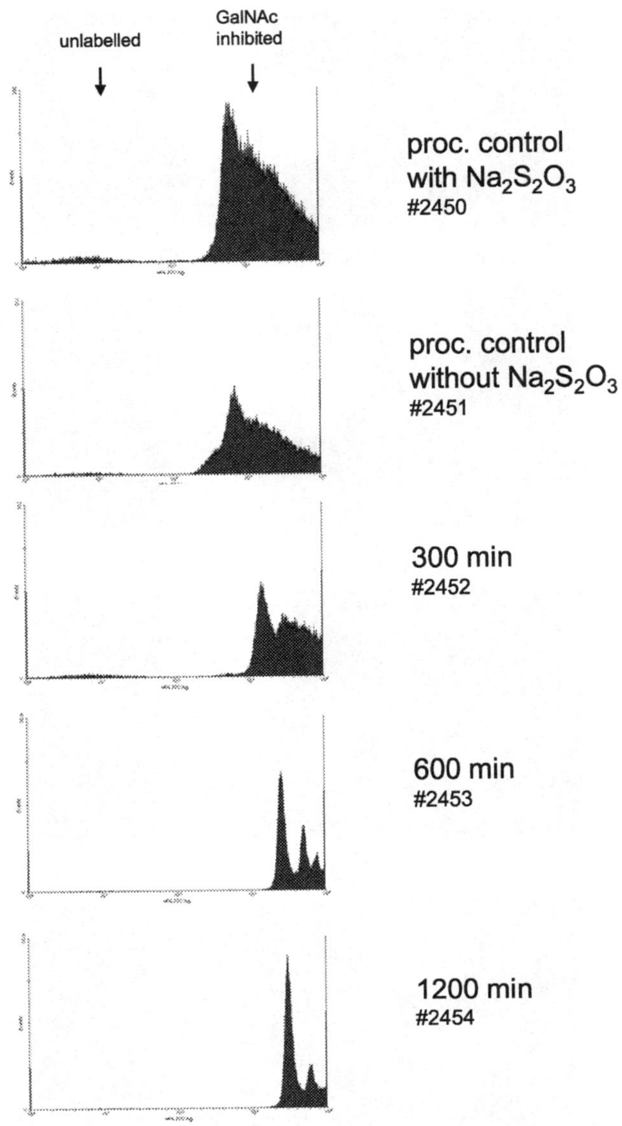

Figure 3.44 Fluorescence histograms of fixed, FITC-MPA labeled GCH1 oocysts. The position of the unlabelled and GalNAc inhibited controls is indicated with vertical arrows. Chlorination conditions were 4 mg/l chlorine, pH 5, 22°C. Two process controls (quenched and unquenched) and 3 chlorine exposures (300, 600 and 1200 min) are shown.

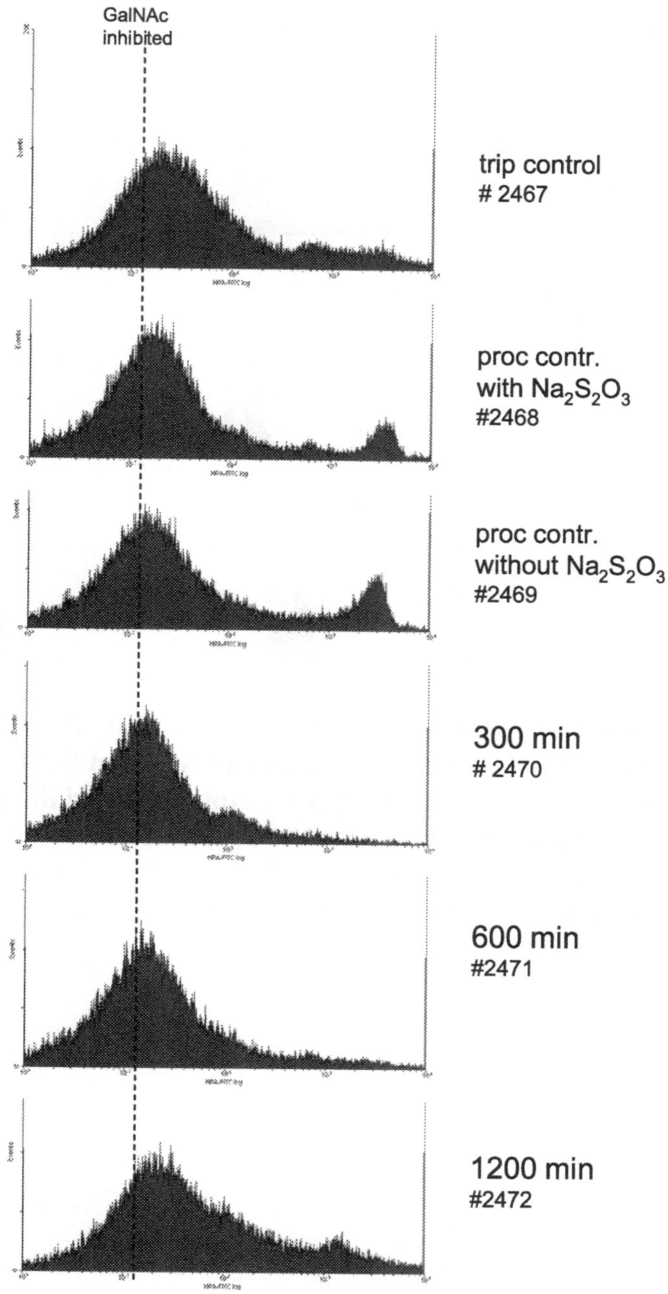

Figure 3.45 Fluorescence histograms of fixed, FITC-HPA labeled GCH1 oocysts. The position of the unlabelled and GalNAc inhibited controls is indicated with vertical arrows. Chlorination conditions were 4 mg/l chlorine, pH 5, 22°C. Two process controls (quenched and unquenched) and 3 chlorine exposures (300, 600 and 1200 min) are shown.

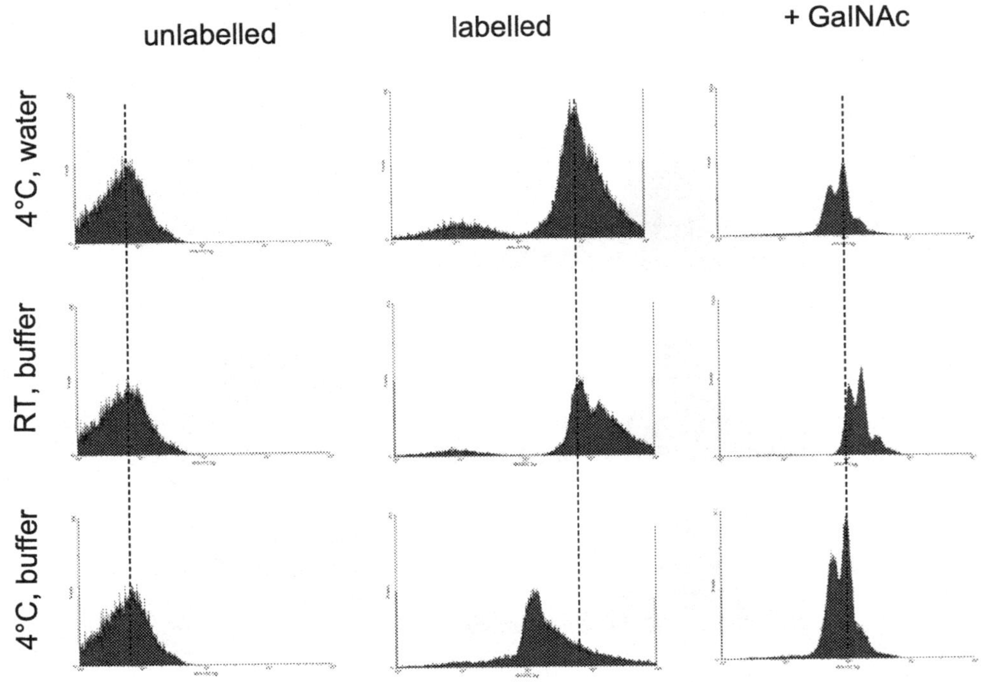

Figure 3.46 Fluorescence histograms of oocysts subjected to three "control treatments." Oocysts were left in water at 4°C, suspended in buffer at room temperature, as used in the process control, and suspended in buffer at 4°C. Note the variability of the fluorescence profiles induced by these treatments. Also note the presence of two fluorescence peaks in the oocysts labeled in the presence of GalNAc. The origin of these peaks was further investigated as described in the text. Vertical dashed lines are shown to facilitate the comparison between samples.

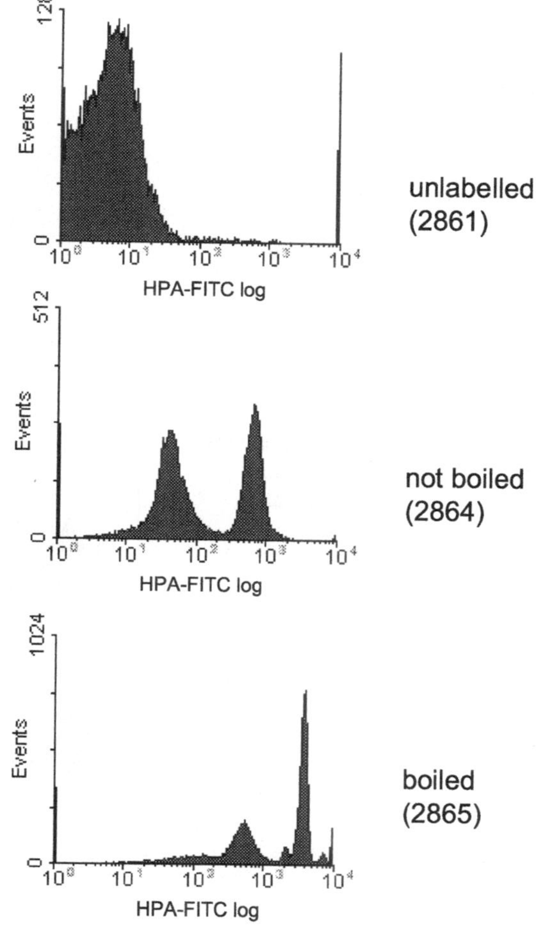

Figure 3.47 Uptake of NBD labeled phosphocholine (NBD-C_6-HPC) into GCH1 oocysts that had been either boiled or unboiled in 1% SDS. Unlabelled oocysts were not exposed to NBD-C_6-HPC but otherwise treated in the same way as unboiled/labeled sample. The graphs are fluorescence histograms, with the fluorescence intensity shown on the x-axis on a log scale and the number of events on the y-axis. Numbers in parentheses are flow cytometric file numbers for internal reference. The frequency distributions of unboiled and boiled oocyst fluorescence indicates that boiling increased dye uptake by 5- to 6-fold.

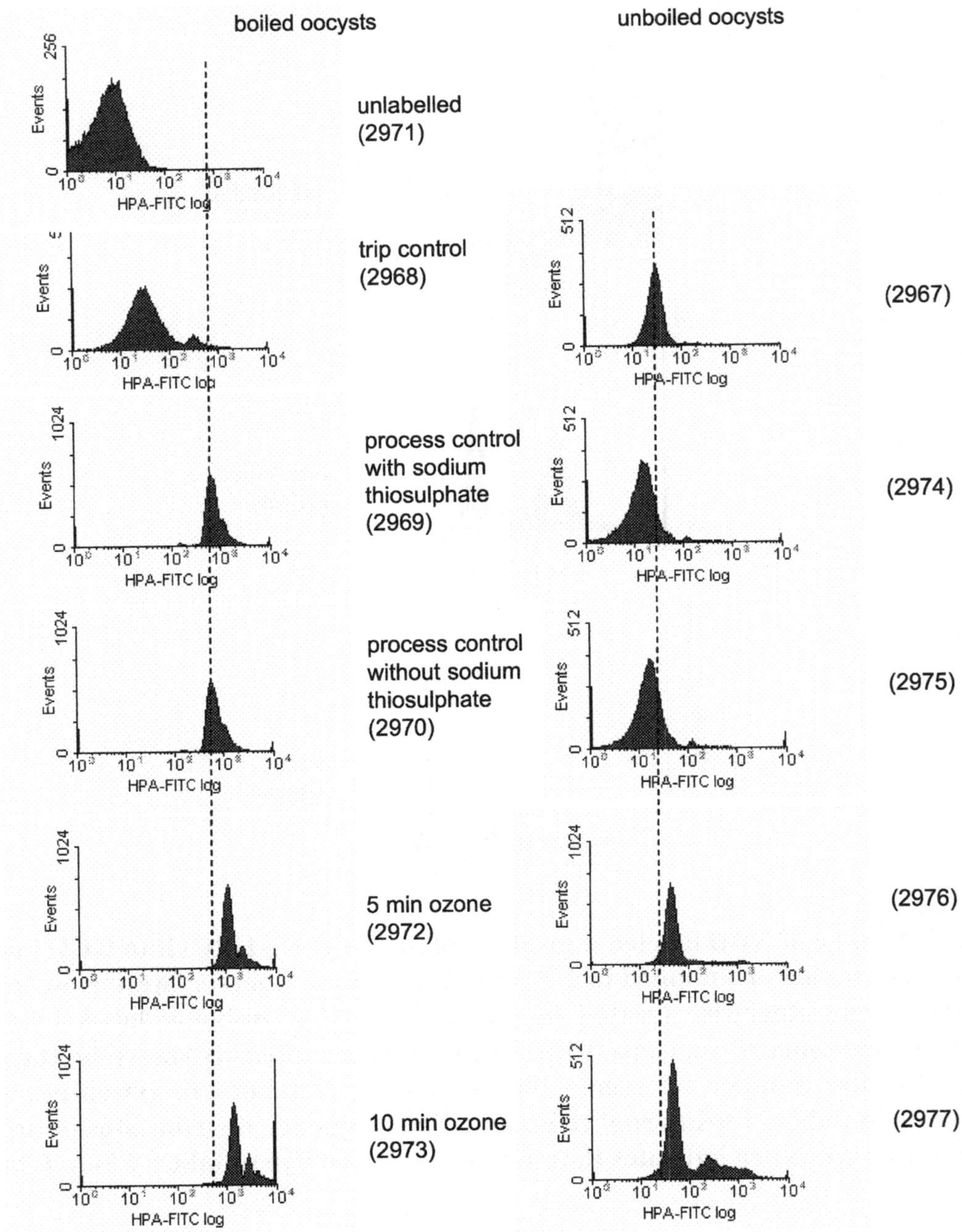

Figure 3.48 Flow cytometric analysis of *C. parvum* genotype II oocysts exposed to ozone. Fluorescence histograms were obtained from ozone treated and control oocysts labeled with NBD-C_6-HPC. Left, boiled; right, unboiled oocysts.

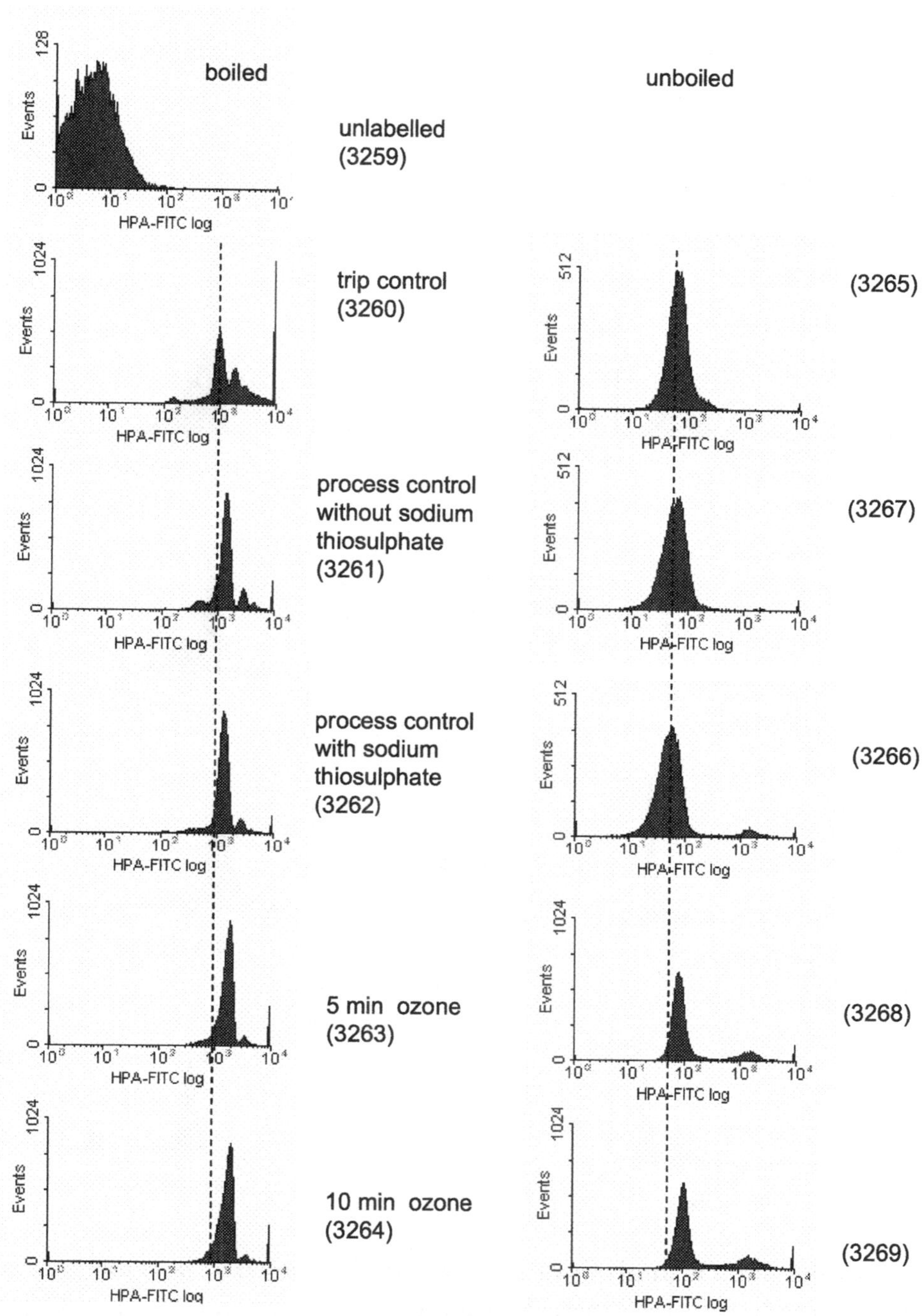

Figure 3.49 Flow cytometric analysis of NBD-C_6-HPC labeled *C. parvum* genotype II oocysts exposed to ozone. See legend to Figure 3.48 for details.

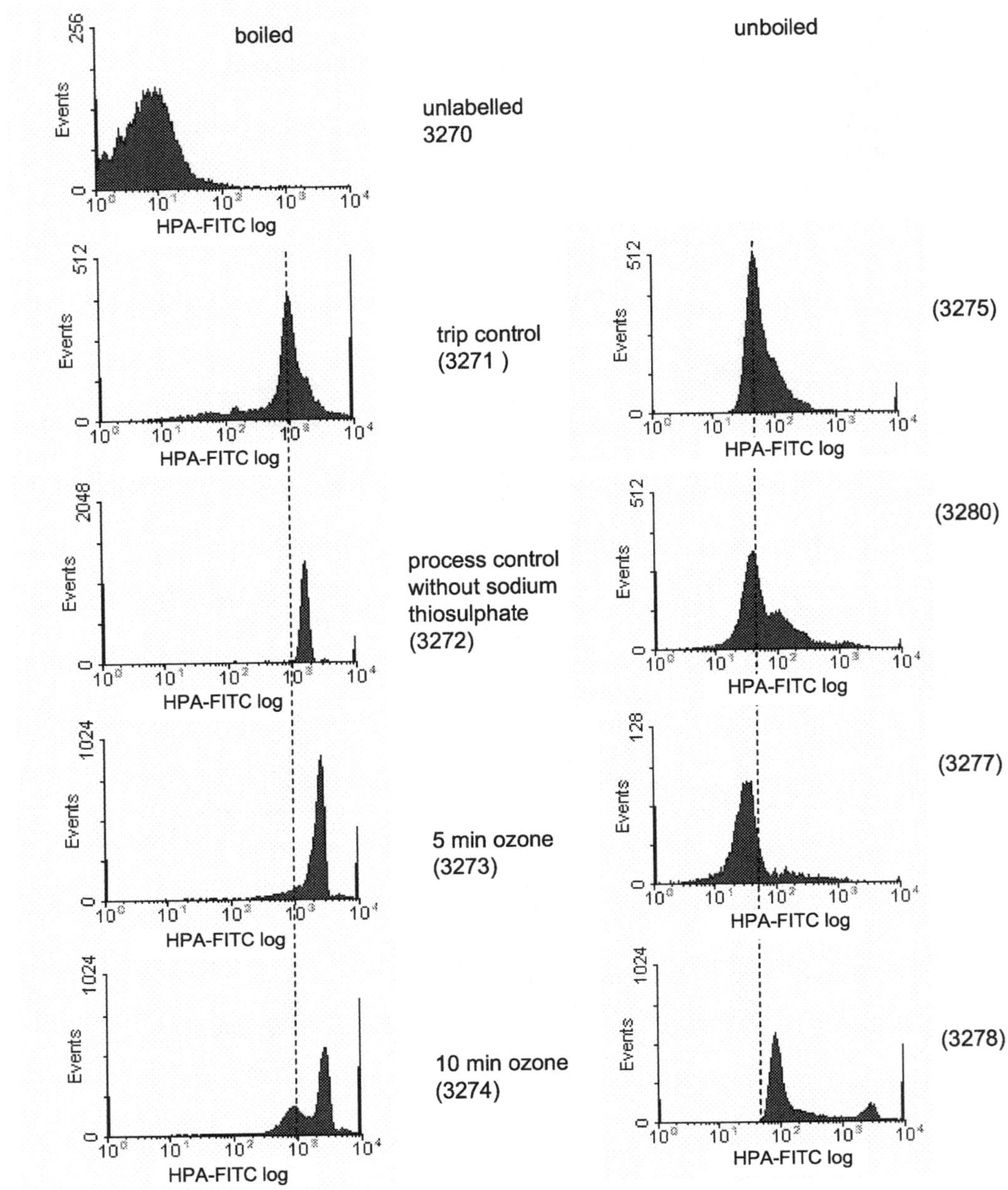

Figure 3.50 Flow cytometric analysis of NBD-C_6-HPC labeled *C. parvum* genotype I oocysts exposed to ozone. See legend to Figure 3.48 for details. Oocysts of isolate TU502 were used in this experiment.

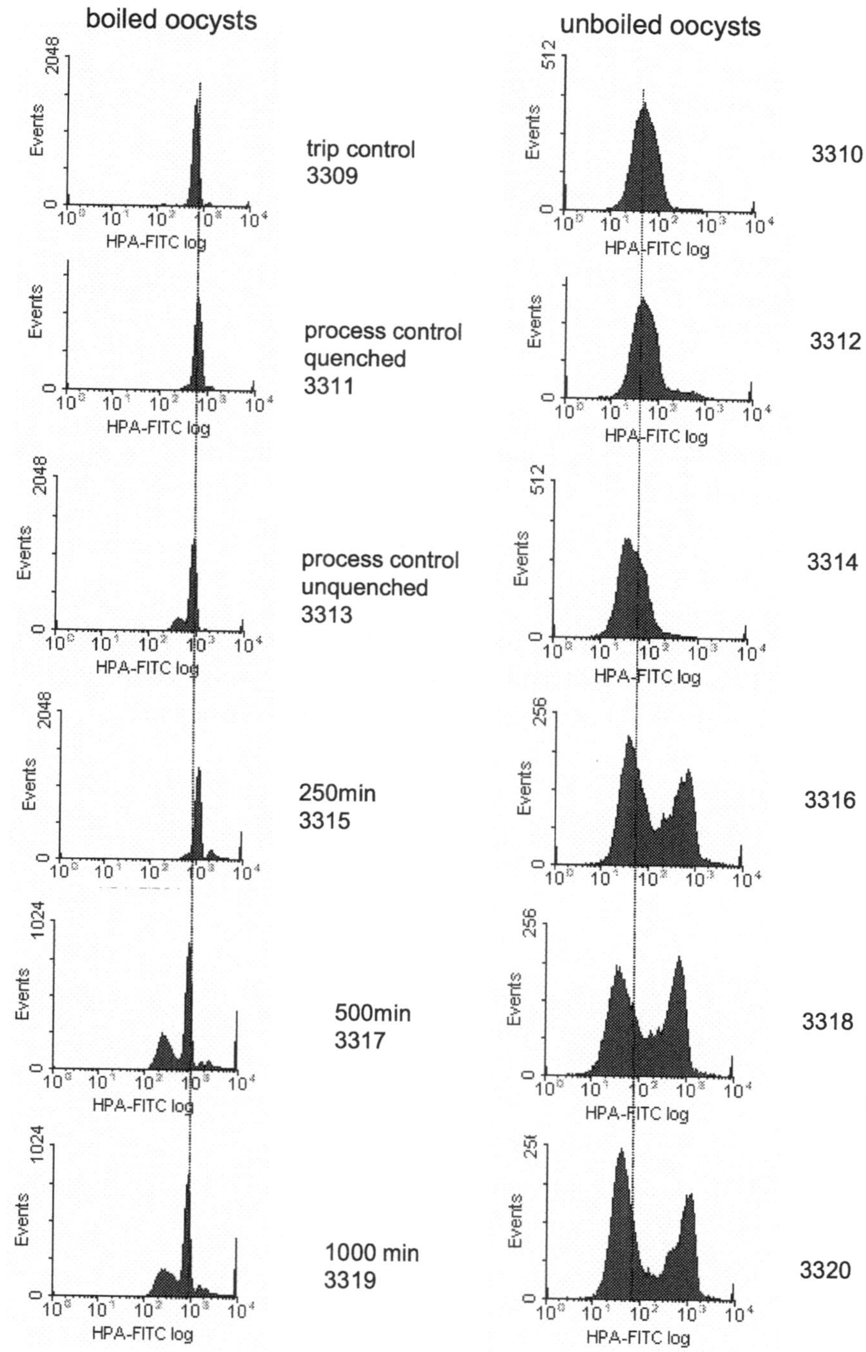

Figure 3.51 Flow cytometric analysis of *C. parvum* genotype II oocysts exposed to chlorine and labeled with NBD-C_6-HPC. Chlorination was performed with 4 mg/l chlorine, pH 5, 22°C.

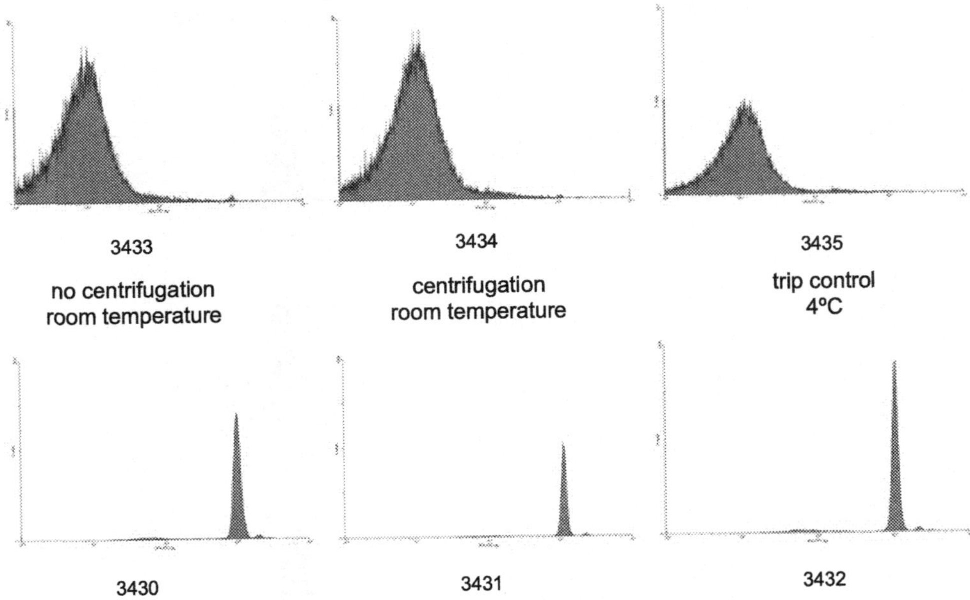

Figure 3.52 Effect of temperature and centrifugation on NBD-C_6-HPC uptake by *C. parvum* genotype II oocysts. Centrifugation conditions were 7000 × g, 10 min at room temperature. Top row, unboiled oocysts; bottom row, boiled oocysts.

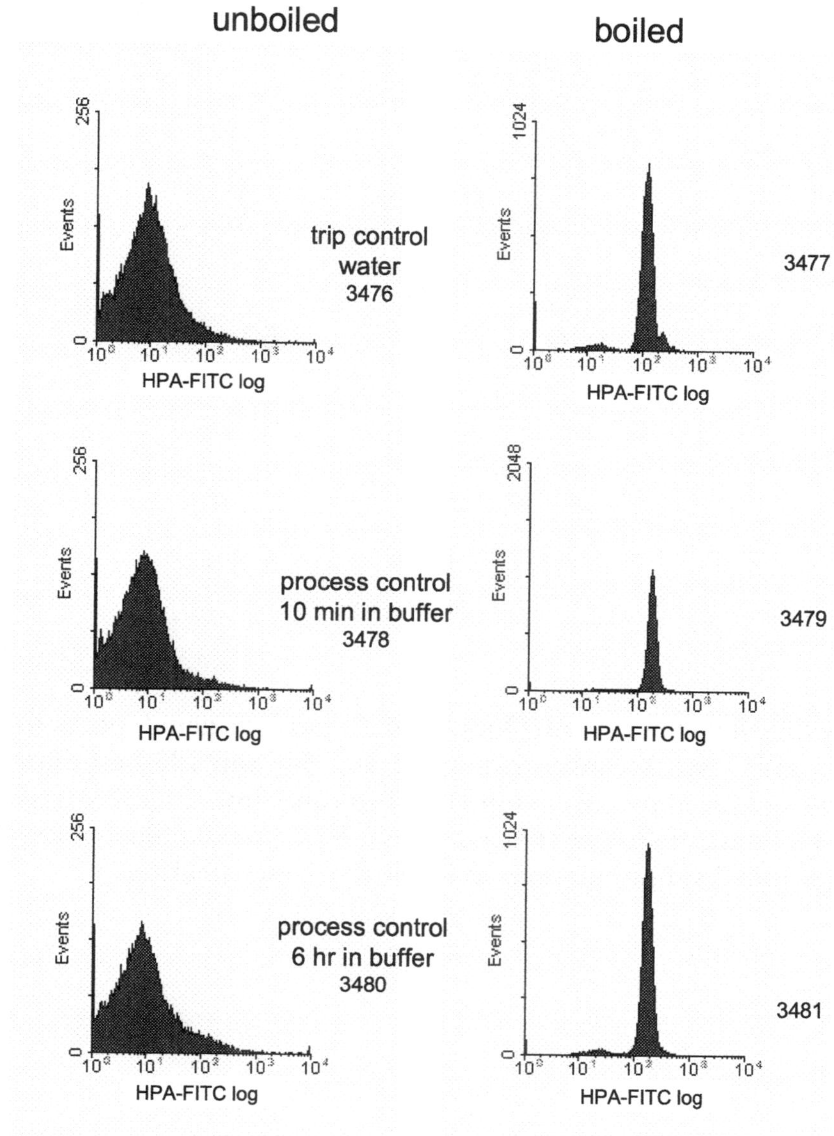

Figure 3.53 Effect of incubation of *C. parvum* genotype II oocyst in phosphate buffer on the uptake of NBD-C_6-HPC. Oocysts were either kept in water at 4°C (trip control conditions), or stirred in phosphate buffer at room temperature (process control conditions) for the time indicated. These treatments had no effect on dye uptake by unboiled oocysts. Boiled process control oocysts showed a minor (<2-fold) increase in dye uptake (right column).

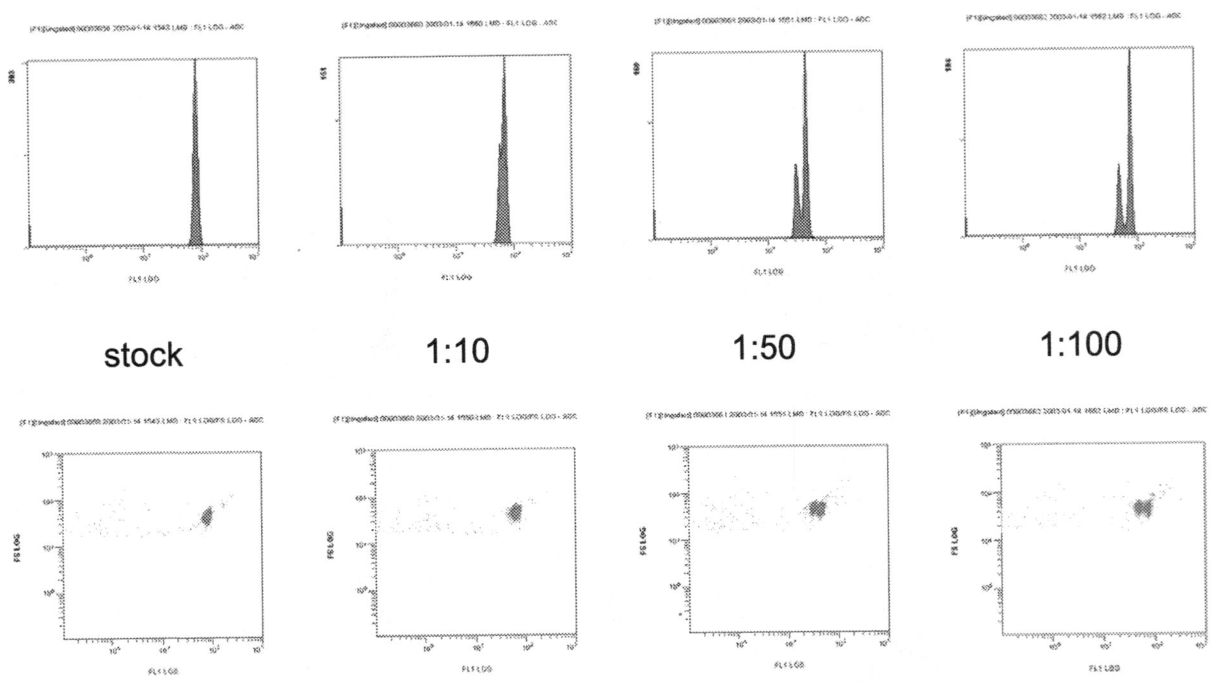

Figure 3.54 Dye dilution experiment. Oocysts of *C. parvum* genotype II were stained with 1 mg/ml NBD-C_6-HPC (stock concentration) or 10-, 50- and 100-fold dilutions of this dye. Fluorescence was measured by flow cytometry. Top row shows FL1 fluorescence histograms. Bottom row shows forward scatter (FS, y-axis) vs. FL1 (x-axis) illustrating the partitioning of the oocyst peak into 2 subpopulations at lower dye concentrations.

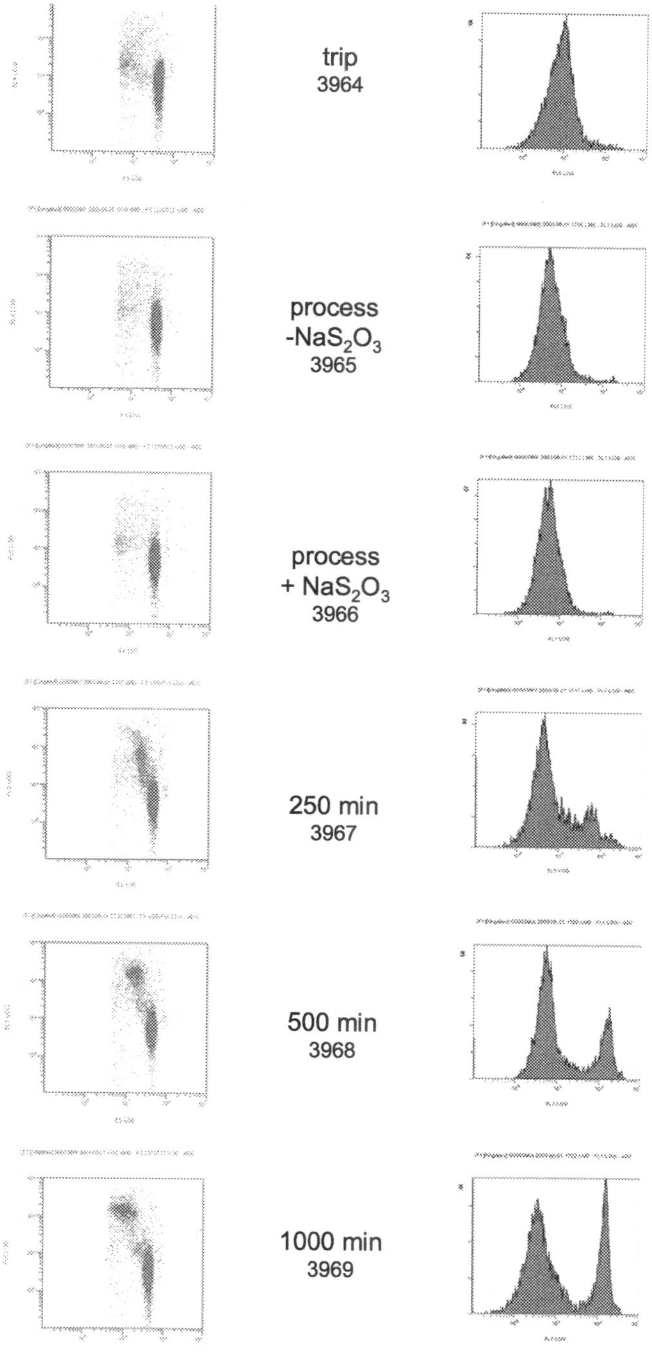

Figure 3.55 Flow cytometric analysis of *C. parvum* genotype II oocysts exposed to chlorine and labeled with NBD-C_6-HPC (repeat experiment). In the left column FL1 fluorescence is plotted on the y-axes and forward scatter on the x-axes. The right column shows FL1 fluorescence histograms. From top to bottom, samples are trip control, process control without NaS_2O_3, process control with NaS_2O_3, 250 min chlorine exposure, 500 min chlorine exposure and 1000 min chlorine exposure. Oocysts were labeled with NBD-C_6-HPC in the presence of 1% SDS.

REFERENCES

Akiyoshi, D.E., X. Feng, M.A. Buckholt, G. Widmer and S. Tzipori. 2002. Genetic Analysis of a *Cryptosporidium parvum* Human Genotype I Isolate Passaged Through Different Host Species. *Infect. Immun.*, 70(10):5670–5.

Anthony, L., D. Bowman, B. Jenkins, S. Eaglesham, K. Kachlany and W.C. Ghiorse. 1997. Abstract Q229. In *97th General Meeting of the American Society for Microbiology.*

Arrowood, M.J., and C.R. Sterling. 1987. Isolation of *Cryptosporidium* Oocysts and Sporozoites Using Discontinuous Sucrose and Isopycnic Percoll Gradients. *J. Parasitol.*, 73(2):314–9.

Bonnin, A., M.N. Fourmaux, J.F. Dubremetz, R.G. Nelson, P. Gobet, G. Harly, M. Buisson, D. Puygauthier-Toubas, G. Gabriel-Pospisil, M. Naciri and P. Camerlynck. 1996. Genotyping Human and Bovine Isolates of *Cryptosporidium parvum* by Polymerase Chain Reaction-Restriction Fragment Length Polymorphism Analysis of a Repetitive DNA Sequence. *FEMS Microbiol. Lett.*, 137(2–3):207–11.

Bozzola, J. 1999. *Electron Microscopy, Principles and Techniques for Biologists.* Jones and Bartlett Publishers.

Campbell, I., A.S. Tzipori, G. Hutchison and K.W. Angus. 1982. Effect of Disinfectants on Survival of *Cryptosporidium* Oocysts. *Vet. Rec.*, 111(18):414–5.

Carraway, M., S. Tzipori, and G. Widmer. 1996. Identification of Genetic Heterogeneity in the *Cryptosporidium parvum* Ribosomal Repeat. *Appl. Environ. Microbiol.*, 62(2):712–6.

Carraway, M., S. Tzipori, and G. Widmer. 1997. A New Restriction Fragment Length Polymorphism From *Cryptosporidium parvum* Identifies Genetically Heterogeneous Parasite Populations and Genotypic Changes Following Transmission From Bovine to Human Hosts. *Infect. Immun.*, 65(9):3958–60.

Cevallos, A.M., N. Bhat, R. Verdon, D.H. Hamer, B. Stein, S. Tzipori, M.E. Pereira, G.T. Keusch and H.D. Ward. 2000. Mediation of *Cryptosporidium parvum* Infection In Vitro by Mucin-Like Glycoproteins Defined by a Neutralizing Monoclonal Antibody. *Infect. Immun.*, 68(9):5167–75.

Corso, P.S., M.H. Kramer, K.A. Blair, D.G. Addiss, J.P. Davis and A.C. Haddix. 2003. Cost of Illness in the 1993 Waterborne *Cryptosporidium* Outbreak, Milwaukee, Wisconsin. *Emerg. Infect. Dis.*, 9(4):426–31.

Current, W.L., and N.C. Reese. 1986. A Comparison of Endogenous Development of Three Isolates of *Cryptosporidium* in Suckling Mice. *J. Protozool.*, 33(1):98–108.

Dalton, D.K., S. Pitts-Meek, S. Keshav, I.S. Figari, A. Bradley and T.A. Stewart. 1993. Multiple Defects of Immune Cell Function in Mice With Disrupted Interferon-Gamma Genes. *Science*, 259(5102):1739–42.

DuPont, H.L., C.L. Chappell, C.R. Sterling, P.C. Okhuysen, J.B. Rose and W. Jakubowski. 1995. The Infectivity of *Cryptosporidium parvum* in Healthy Volunteers. *New Engl. J. Med.*, 332(13):855–9.

Fayer, R., U. Morgan and S.J. Upton. 2000. Epidemiology of *Cryptosporidium*: Transmission, Detection and Identification. *Int. J. Parasitol.*, 30(12–13):1305–22.

Fricker, C.R., and J.H. Crabb. 1998. Water-borne Cryptosporidiosis: Detection Methods and Treatment Options. *Adv. Parasitol.*, 40:241–78.

Goding, J. 1986. *Monoclonal Antibodies, Principles and Practice.* London: Academic Press.

Griffiths, J.K. 1998. Human Cryptosporidiosis: Epidemiology, Transmission, Clinical Disease, Treatment, and Diagnosis. *Adv. Parasitol.*, 40:37–85.

Griffiths, J.K., C. Theodos, M. Paris and S. Tzipori. 1998. The Gamma Interferon Gene Knockout Mouse: A Highly Sensitive Model for Evaluation of Therapeutic Agents Against *Cryptosporidium parvum*. *J. Clin. Microbiol.*, 36(9):2503–8.

Guerrant, R.L. 1997. Cryptosporidiosis: An Emerging, Highly Infectious Threat. *Emerg. Infect. Dis.*, 3(1):51–7.

Gut, J., and R.G. Nelson. 1994. *Cryptosporidium parvum* Sporozoites Deposit Trails of 11A5 Antigen During Gliding Locomotion and Shed 11A5 Antigen During Invasion of MDCK Cells In Vitro. *J. Eukaryot. Microbiol.*, 41(5):42S–43S.

Harlow, E., and D. Lane. 1988. *Antibodies: A Laboratory Manual.* Plainview, N.Y.: Cold Spring Harbor Laboratory Press.

Korich, D.G., J.R. Mead, M.S. Madore, N.A. Sinclair and C.R. Sterling. 1990. Effects of Ozone, Chlorine Dioxide, Chlorine, and Monochloramine on *Cryptosporidium parvum* Oocyst Viability. *Appl. Environ. Microbiol.*, 56(5):1423–8.

Lally, N.C., G.D. Baird, S.J. McQuay, F. Wright and J.J. Oliver. 1992. A 2359-base Pair DNA Fragment From *Cryptosporidium parvum* Encoding a Repetitive Oocyst Protein. *Mol. Biochem. Parasitol.*, 56(1):69–78.

Leav, B.A., M. Mackay and H.D. Ward. 2003. *Cryptosporidium* Species: New Insights and Old Challenges. *Clin. Infect. Dis.*, 36(7):903–8.

Llovo, J., A. Lopez, J. Fabregas and A. Munoz. 1993. Interaction of Lectins With *Cryptosporidium parvum*. *J. Infect. Dis.*, 167(6):1477–80.

Lumb, R., J.A. Lanser and P.J. O'Donoghue. 1988. Electrophoretic and Immunoblot Analysis of *Cryptosporidium* Oocysts. *Immunol. Cell Biol.*, 66 (Pt 5–6):369–76.

MacKenzie, W.R., N.J. Hoxie, M.E. Proctor, M.S. Gradus, K.A. Blair, D.E. Peterson, J.J. Kazmierczak, D.G. Addiss, K.R. Fox, J.B. Rose and et al. 1994. A Massive Outbreak in Milwaukee of *Cryptosporidium* Infection Transmitted Through the Public Water Supply. *New Engl. J. Med.*, 331(3):161–7.

Manzi, A.E., and H.V. Halbeek. 1999. *Essentials of Glycobiology.* Plainview, N.Y.: Cold Spring Harbor Laboratory Press.

Mead, J.R., R.C. Humphreys, D.W. Sammons and C.R. Sterling. 1990. Identification of Isolate-Specific Sporozoite Proteins of *Cryptosporidium parvum* by Two-Dimensional Gel Electrophoresis. *Infect. Immun.*, 58(7):2071–5.

Mitschler, R.R., R. Welti and S.J. Upton. 1994. A Comparative Study of Lipid Compositions of *Cryptosporidium parvum* (Apicomplexa) and Madin-Darby Bovine Kidney Cells. *J. Eukaryot. Microbiol.*, 41(1):8–12.

Morgan, U.M., C.C. Constantine, D.A. Forbes and R.C. Thompson. 1997. Differentiation Between Human and Animal Isolates of *Cryptosporidium parvum* Using rDNA Sequencing and Direct PCR Analysis. *J. Parasitol.*, 83(5):825–30.

Morgan, U.M., C.C. Constantine, P. O'Donoghue, B.P. Meloni, P.A. O'Brien and R.C. Thompson. 1995. Molecular Characterization of *Cryptosporidium* Isolates From Humans and Other Animals Using Random Amplified Polymorphic DNA Analysis. *Am. J. Trop. Med. Hyg.*, 52(6):559–64.

Nanduri, J., S. Williams, T. Aji and T.P. Flanigan. 1999. Characterization of an Immunogenic Glycocalyx on the Surfaces of *Cryptosporidium parvum* Oocysts and Sporozoites. *Infect. Immun.*, 67(4):2022–4.

O'Donoghue, P.J. 1995. *Cryptosporidium* and Cryptosporidiosis in Man and Animals. *Int. J. Parasitol.*, 25(2):139–95.

Okhuysen, P.C., S.M. Rich, C.L. Chappell, K.A. Grimes, G. Widmer, X. Feng and S. Tzipori. 2002. Infectivity of a *Cryptosporidium parvum* Isolate of Cervine Origin for Healthy Adults and Interferon-Gamma Knockout Mice. *J. Infect. Dis.*, 185(9):1320–5.

Ortega, Y.R., R.R. Sheehy, V.A. Cama, K.K. Oishi and C.R. Sterling. 1991. Restriction Fragment Length Polymorphism Analysis of *Cryptosporidium parvum* Isolates of Bovine and Human Origin. *J. Protozool.*, 38(6):40S–41S.

Peng, M.M., L. Xiao, A.R. Freeman, M.J. Arrowood, A.A. Escalante, A.C. Weltman, C.S. Ong, W.R. Mac Kenzie, A.A. Lal and C.B. Beard. 1997. Genetic Polymorphism Among *Cryptosporidium parvum* Isolates: Evidence of Two Distinct Human Transmission Cycles. *Emerg. Infect. Dis.*, 3(4):567–73.

Ranucci, L., H.M. Muller, G. La Rosa, I. Reckmann, M.A. Morales, F. Spano, E. Pozio and A. Crisanti. 1993. Characterization and Immunolocalization of a *Cryptosporidium* Protein Containing Repeated Amino Acid Motifs. *Infect. Immun.*, 61(6):2347–56.

Reese, N.C., W.L. Current, J.V. Ernst and W.S. Bailey. 1982. Cryptosporidiosis of Man and Calf: A Case Report and Results of Experimental Infections in Mice and Rats. *Am. J. Trop. Med. Hyg.*, 31(2):226–9.

Rose, J., T. Lisle and M. LeChevallier. 1997. Waterborne Cryptosporidiosis: Incidence, Outbreak and Treatment Strategies. In Cryptosporidium *and Cryptosporidiosis*. Edited by R. Fayer. New York: CRC Press. pp. 93–110.

Sambrook, J., and D.W. Russell. 2001. *Molecular Cloning: A Laboratory Manual*. Plainview, N.Y.: Cold Spring Harbor Laboratory Press.

Spano, F., C. Puri, L. Ranucci, L. Putignani and A. Crisanti. 1997a. Cloning of the Entire COWP Gene of *Cryptosporidium parvum* and Ultrastructural Localization of the Protein During Sexual Parasite Development. *Parasitol.*, 114(Pt 5):427–37.

Spano, F., L. Putignani, J. McLauchlin, D.P. Casemore and A. Crisanti 1997b. PCR-RFLP Analysis of the *Cryptosporidium* Oocyst Wall Protein (COWP) Gene Discriminates Between *C. wrairi* and *C. parvum*, and Between *C. parvum* Isolates of Human and Animal Origin. *FEMS Microbiol. Lett.*, 150(2):209–17.

Spano, F., L. Putignani, A. Crisanti, P. Sallicandro, U.M. Morgan, S.M. Le Blancq, L. Tchack, S. Tzipori and G. Widmer. 1998. Multilocus Genotypic Analysis of *Cryptosporidium parvum* Isolates From Different Hosts and Geographical Origins. *J. Clin. Microbiol.*, 36(11):3255–9.

Theodos, C.M., K.L. Sullivan, J.K. Griffiths and S. Tzipori. 1997. Profiles of Healing and Nonhealing *Cryptosporidium parvum* Infection in C57BL/6 Mice With Functional B and T Lymphocytes: The Extent of Gamma Interferon Modulation Determines the Outcome of Infection. *Infect. Immun.*, 65(11):4761–9.

Tilley, M., and S.J. Upton. 1997. Biochemistry of *Cryptosporidium*. In Cryptosporidium *and Cryptosporidiosis*. Edited by R. Fayer. New York: CRC Press. pp. 163–180.

Tilley, M., S.J. Upton, B.L. Blagburn and B.C. Anderson. 1990. Identification of Outer Oocyst Wall Proteins of Three *Cryptosporidium* (Apicomplexa: Cryptosporidiidae) Species by 125I Surface Labeling. *Infect. Immun.*, 58(1):252–3.

Tzipori, S., and J.K. Griffiths. 1998. Natural History and Biology of *Cryptosporidium parvum*. *Adv. Parasitol.*, 40:5–36.

Tzipori, S., W. Rand, J. Griffiths, G. Widmer and J. Crabb. 1994. Evaluation of an Animal Model System for Cryptosporidiosis: Therapeutic Efficacy of Paromomycin and Hyperimmune Bovine Colostrum-Immunoglobulin. *Clin. Diag. Lab. Immunol.*, 1(4):450–63.

Tzipori, S., and G. Widmer. 2000. The Biology of *Cryptosporidium. Contrib. Microbiol.*, 6:1–32.

Ungar, B. 1990. Cryptosporidiosis in Humans (*Homo sapiens*). In *Cryptosporidiosis of Man and Animals.* Edited by J. Dubey, C. Speer, and R. Fayer. New York: CRC Press. pp. 59–82.

Widmer, G., D. Akiyoshi, M.A. Buckholt, X. Feng, S.M. Rich, K.M. Deary, C.A. Bowman, P. Xu, Y. Wang, X. Wang, G.A. Buck and S. Tzipori. 2000. Animal Propagation and Genomic Survey of a Genotype I Isolate of *Cryptosporidium parvum. Mol. Biochem. Parasitol.*, 108(2):187–97.

Widmer, G., T. Clancy, H.D. Ward, D. Miller, G.M. Batzer, C.B. Pearson and Z. Bukhari. 2002. Structural and Biochemical Alterations in *Giardia lamblia* Cysts Exposed to Ozone. *J. Parasitol.*, 88(6):1100–6.

Widmer, G., S. Tzipori, C.J. Fichtenbaum and J.K. Griffiths. 1998. Genotypic and Phenotypic Characterization of *Cryptosporidium parvum* Isolates From People With AIDS. *J. Infect. Dis.*, 178(3):834–40.

ABBREVIATIONS

AIA	*Artocarpus integrifolia* agglutinin
anti-cpRL3	antisera to polypeptide encoded by cpRL3 (see below), generated in rabbits by Dr. Joseph Crabb's laboratory, ImmuCell Corporation
AOA	2-(9-anthroyloxy) oleic acid
BCIP	5-bromo-4-chloro-3-indolyl phosphate
cDNA	complementary deoxyribonucleic acid
Cellabs mAb	IgG_1 monoclonal antibody to *Cryptosporidium* oocysts generated by Cellabs, Australia, and provided to Dr. Honorine Ward, Tufts-New England Medical Center, by Dr. Tony Smithyman
Con A	*Canavalia ensiformis* (Concanavalin A) agglutinin
COWP	*Cryptosporidium* oocyst wall protein (first described by Ranucci et al. 1993)
cpRL3	DNA clone (first isolated by Ranucci et al. 1993, and provided to Dr. Joseph Crabb, ImmuCell Corporation) encoding a polypeptide constituting the amino terminal 786 amino acids of *Cryptosporidium* oocyst wall protein -190 (COWP-190) and ligated into pUA1/pDS56/RBSII-E^-, a 6 × his construct
CrA2	*Cryptosporidium parvum* IgA monoclonal antibody 2, originally generated by Drs. Marian Neutra and Saul Tzipori, Tufts University School of Veterinary Medicine, and provided to Dr. Honorine Ward, Tufts-New England Medical Center
DAPI	4',6'-diamidino-2-phenylindole
DME-PSG	Dulbecco's modified Eagle's medium supplemented with penicillin, streptomycin, and glutamine supplement
DNA	deoxyribonucleic acid
DTT	dithiothreitol
EcoR1	a restriction endonuclease
8A9.2	IgG_1 monoclonal antibody to *C. parvum* genotype I oocysts generated in Saul Tzipori's laboratory, Tufts University School of Veterinary Medicine
FCM	flow cytometry
FITC	fluorescein isothiocyanate
4E9	IgM monoclonal antibody to *C. parvum* zoites generated in Honorine Ward's laboratory, Tufts-New England Medical Center
FPLC	fast protein liquid chromatography
Fuc	fucose

Gal	galactose
GalNAc	N-acetylgalactosamine
GCH1	Grafton *Cryptosporidium* Human 1. An oocyst isolate of genotype II, originally obtained from a human AIDS patient and passage in neonatal calves at Tufts University School of Veterinary Medicine
GC-MS	gas chromatography-mass spectrometry
Glc	glucose
GlcNAc	N-acetylglucosamine
HPA	*Helix pomatia* agglutinin
HPAEC	high pH anion exchange chromatography
HPAEC-PAD	high pH anion exchange chromatography with pulsed amperometric detection
HRP	horseradish peroxidase
ICC	ImmuCell Corporation (Portland, Maine)
IEF	isoelectric focusing
IEM	immunoelectron microscopy
IFA	immunofluorescence assay
IgA	immunoglobulin A (alpha heavy chain)
IGEPAL CA-630	a nonionic detergent
IgG	immunoglobulin G (gamma heavy chain)
IgM	immunoglobulin M (mu heavy chain)
IMS	immunomagnetic separation
λgt11	expression vector for DNA library construction
λzap express	expression vector for DNA library construction (Stratagene)
λzap II	expression vector for DNA library construction (Stratagene)
M	molar
MAA	*Maakia amurennis* agglutinin
mAb	monoclonal antibody
MALDI-TOF MS	matrix-assisted laser desorption/ionization time-of-flight mass spectrometry
Man	mannose
MD	genotype II isolate
MDBK	Madin-Darby Bovine Kidney
MeOH	methanol
mg	milligram
ml	milliliter
mM	millimolar
MPA	*Maclura pomifera* agglutinin
MQW	Milli-Q water
M_r	relative molecular mass
μg	microgram

µl	microliter
µM	micromolar
Na acetate	sodium acetate
NaCl	sodium chloride
NBD	nitrobenzoxadiazolyl
NBD-C_6-HPC	2-(nitrobenzoxadiazolyl-amino)hexanoyl-1-hexadecanoyl-sn-glycero-3-phosphocholine (phospholipid)
NBD-C_6-sphingomyelin	6-(N-(7-nitrobenz-2-oxa-1,3,diazol-4-yl)aminohexanolyl)sphingo-yl phosphocholine
NBD-C_{12}-HPC	2-(nitrobenzoxadiazolyl-amino)dodecanoyl-1-hexadecanoyl-sn-glycero-3-phosphocholine
NBD-Sterol	22-(N-(7-nitrobenz-2-oxa-1,3-dazol-4-yl)amino)-23,24-bisnor-5-cholenol-3b-ol
NBT	nitroblue tetrazolium
NEMC	New England Medical Center (Boston, Mass.)
NEMC1	New England Medical Center 1. An oocyst isolate of genotype I, originally obtained from a human AIDS patient at New England Medical Center and passage in gnotobiotic piglets at Tufts University School of Veterinary Medicine, North Grafton, Mass.
NINC	An oocyst isolate of genotype II, originally obtained from naturally infected neonatal calves
nM	nanomolar
1D	one-dimensional
PAGE	polyacrylamide gel electrophoresis
PBS	phosphate-buffered saline
PF	paraformaldehyde
PNA	*Arachis hypogea* (peanut) agglutinin
Sau3A1	a restriction endonuclease
SBA	*Glycine max* (soybean) agglutinin
SDS	sodium dodecyl sulfate
SJA	*Sophora japanicum* agglutinin
SNA	*Sambucus nigrans* agglutinin
TEM	transmission electron microscopy
TMB	3,3',5,5'-tetramethylbenzidine
TU502	An oocyst isolate of genotype I (originally called UG502), originally obtained from a human patient in Uganda and passaged in gnotobiotic piglets and neonatal calves at Tufts University School of Veterinary Medicine
TUSVM	Tufts University School of Veterinary Medicine (North Grafton, Mass.)

2C3 mAb	IgG_3 monoclonal antibody to *Cryptosporidium* oocysts generated by Joe Crabb's laboratory, ImmuCell Corporation
2D	two-dimensional
2-ME	2-β-mercaptoethanol
UCP	An oocyst isolate of genotype II, originally obtained from Dr. Beth Ungar at USUHS isolated from a calf, and passaged in neonatal calves since 1989 at least every 4–6 months under ImmuCell supervision
UCSD	University of California, San Diego
UCSF	University of California, San Francisco
UEA-1	*Ulex europeus* agglutinin-1
UG502	Uganda 502. An oocyst isolate of genotype I, originally obtained from a human patient in Uganda and passage in gnotobiotic piglets and neonatal calves at Tufts University School of Veterinary Medicine
USUHS	Uniformed Services University of the Health Sciences
VVA	*Vicia villosa* agglutinin
WGA	*Triticum vulgare* (wheat germ) agglutinin

IWA Publishing's authorised EU representative for General Product Safety Regulations is Diane D'Arras, 15 rue Duret, 75116 Paris, France, e-mail: safety@iwap.co.uk.

Printed and bound by CPI Group (UK) Ltd, Croydon, CR0 4YY
17/09/2026
02218802-0013